A HISTORY OF IRISH STEEL

A HISTORY OF IRISH STEEL

Sarsfield Hogan

GILL AND MACMILLAN

First published 1980 by
Gill and Macmillan Ltd
15/17 Eden Quay
Dublin 1
with associated companies in
London, New York, Delhi, Hong Kong,
Johannesburg, Lagos, Melbourne,
Singapore, Tokyo

0 7171 1025 7

Origination by Healyset, Dublin
Printed and bound in Ireland by
Cahill Printers Limited

Contents

Dedicated to
the men and women
of the Irish steel industry

List of Illustrations

List of illustrations continued

Foreword by an Taoiseach, Jack Lynch, B.L., T.D.

Nobody is better qualified than Sarsfield Hogan to write the story of Irish Steel. He was one of the three civil servants to whom the task was given by Sean Lemass in 1947 to bring the steel works in Haulbowline into production and he remained with the company for the following twenty-seven years. As a director at first, as chairman from June 1956 and as executive chairman from 1961, he steered the industry through many vicissitudes and, since passing the baton to his successor, he has found time to record in a very readable way the chronicle of steelmaking at Haulbowline. His book describes the development of the industry under state sponsorship after the first endeavours under private enterprise were frustrated by the onset of the Second World War.

It is appropriate that this book should be published when the company is embarking on an important new development programme which will double its capacity and ensure that it will be competitive in the free-trade environment of today.

While thanking Sarsfield Hogan for his past service and for this absorbing book, may I also wish the company and all its employees a most successful future.

Jack Lynch

Taoiseach, November 1979

Preface

Good God! What have I taken on me to do? Truly I am an ass, otherwise I would never have undertaken so heavy a burthen: but no matter, I shall do what I can, like Coppinger's female dog, who always took her own way.

Captain Bodley's account of his journey into Lecale in Ulster at Christmas 1602.

These opening words from the gallant Captain's manuscript (which is preserved in the British Museum and excerpted in *The Old Irish World* by Mrs A. S. Green, 1912) made some appeal as touching my own situation in undertaking to write this book.

My path follows the first forty years of steel manufacture in Ireland, more precisely on Haulbowline Island in Cork Harbour. The story is, I am encouraged to believe, worth recording as part of Irish industrial history – but does it sustain enough interest to justify publication? The answer may be found in the generally accepted importance – indeed, necessity – of a steel industry in any developed economy. Such an industry came to Ireland at a remarkably late date and, after many vicissitudes, is still at the stage of consolidation and reconstruction. The rolling of merchant steel from imported material began at Haulbowline in August 1939 during most unfavourable international circumstances. The first furnace production of steel from native scrap took place seven years later and has continued since then with much improvement over the years in equipment and melting shop practices. The further step of direct reduction of iron ore, or its equivalent in pellet form, into metallic charge for steel-making furnaces has yet to come. It will be a big step requiring much capital, competitive costings and assured external markets. The same requirements govern the important project now in train for expanding and modernising the Haulbowline plant so as to secure its future in the European Communities.

Thus the story being told is not yet complete. It is not one of triumphant success or spectacular achievements but rather a record, not indeed of defeat, but of persistent endeavour, struggle and survival, interspersed with sunnier days as well as periods of acute crisis. In a note written to Sean McEntee, Minister for Finance, in 1952, I observed that five years with the steel industry had taught those in charge of it to avoid complacency, because when one major problem had been surmounted, another inevitably reared its head. The pattern has not changed during the subsequent quarter of a century. While such tribulations have sometimes been of domestic origin, more often they have stemmed from world-wide movements affecting the steel industry. The cyclical waves of boom and depression have tormented planners and managers in the great works of South Wales, in the Ruhr and along the Ohio Valley. Those striving on a far humbler scale within Cork Harbour suffered in good company, although this was not much consolation.

It can be claimed that the story of the Irish steel industry is an unusual episode in the country's economic fortunes. Successive Irish governments took measures to promote industry, accepted responsibility for the running of public utilities such as power and transport and, in some instances, assumed ownership of industrial operations (as in the case of sugar) in pursuance of a well-defined policy. The involvement in steel production, however, was of a different kind. It came about by accident rather than design. No political party or Minister had ever declared 'Ireland must have a steel industry and the government will establish one'. Like a foundling orphan, the desolate steel works at Haulbowline was taken under public care and protection in 1947. The child was nurtured kindly with the least possible fuss and public expense for more than a decade, eventually achieving some degree of legitimacy as an object of state policy in the First Programme of Economic Expansion. Twenty years later, one surveys the adult offspring of public ownership, a strong industrial enclave in human terms but beset with problems in a changing world and calling for greater investment to secure its future.

Another aspect of the story, which is of interest to the student of Irish public affairs, concerns the unusual methods adopted in giving effect to the decision taken in 1947 to conduct steel-making as a state enterprise. It was a move of expediency from which economic conviction was absent. The normal procedure when state undertakings were established was to introduce legislation which contained all the necessary provisions for organisation, financing and public accountability. This can be observed in the Electricity Supply Acts, the Air Navigation and Transport Acts, the Turf Development Acts, to name but a few instances. In the case of the steel industry, the measures necessary were implemented by administrative action without any formal parliamentary approval. Irish Steel Holdings, Ltd. was incorporated as a private company by three civil servants under ministerial authority and direction. It was not until 1960 that legislation became essential in order to enable the Minister for Finance to buy and sell shares in the company. Even today the student must seek out the Memorandum and Articles of Association of the company when he wishes to ascertain its objectives and the degree of public control under which it fulfils its manufacturing and trading activities. The informal and non-statutory arrangements which governed its existence were no handicap to the company. Indeed they gave it room for flexibility and opportunism which were, at times, valuable aids to commercial progress in an unfamiliar and exacting market place.

I am conscious that this book must suffer from the fact that the historian played a part in many of the events which are discussed. It is impossible for me to be completely objective in writing about an industry with which I was directly concerned in one role or another for twenty-seven years. Even when I relate the first attempt by private industrialists and (as it seems to me) the elements of dramatic irony, at least in human terms, which attended that ill-starred venture, I am, once again, struggling to find objectivity. Not alone was I acquainted with many of the actors on that stage, but I was also involved as a public servant when eventually the government had to heed the distress signals from Haulbowline.

Thus, it is as well for me to admit at once my shortcomings as an impartial judge. In case of misunderstanding I should add that any comment or opinion expressed or inferred throughout the book, which is not clearly attributed to someone else, is mine alone and is not to be laid at the door of the steel company or of anyone connected with it.

What I have written will be largely concerned with events and people. There is no room, even if I were at all competent, for a description of the technology of steel-making and of the detailed manufacturing processes at Haulbowline. Some technical references do, indeed, occur but these are incidental to certain parts of the story, particularly those concerning development of the industry. I have also avoided discussion, in depth at any rate, of the economic argument for an Irish steel industry and the cost/benefit analysis of the case for state investment. This is not an academic treatise, but a human story of men tried by fortune.

The first chapter of the book may appear to depart from the limitations implied by the preceding paragraph. This piece of pre-history, for such it may be termed in the context of the Irish steel industry, springs from the question the author asked himself when he first approached his task, 'Why no Irish steel before 1939?' I make no claim to answer the question comprehensively. Far from it; indeed, hands other than mine, willing to undertake deeper research, may be stimulated to examine further the curious anomaly in our economic development which paucity of natural resources for steel-making does not explain, however much it provided the cause for discouragement and delay.

I am greatly indebted to my former colleagues in Irish Steel Holdings for their assistance in the compilation of this history. Many of them, including G. Farren, C. P. Vaughan, G. Radley, P. Mortimer, R. H. Warbrook and the late John Rafferty, gave time to helping me. I have, I fear, been a great nuisance to them and others in the company.

I am grateful to the Departments of Finance and Industry and Commerce for allowing me to inspect some of their files of the period 1937–47 so that I could check my own recollection of events and dates. The facilities given me by the Public Record Office, the National Library and the Royal Irish Academy have been invaluable.

My thanks are due to many individuals who have been so willing to help me with information or recollections or suggestions. I am indebted to Professor Patrick Lynch, who kindly read my early manuscript and gave me such encouraging comments and also to Dr Brendan Menton, David Duff Frame, James Davy, Aodhagán O'Rahilly, Alf Donohoe, Christopher Fitzpatrick, Pádraig Ó Maidín and Niall Brunicardi.

The main burden of copying my manuscripts has fallen on Mrs Germaine Dolan's Agency; to her and her staff my thanks are due for unfailing efficiency. In the same field of essential back-up for my efforts, I am indebted to Mrs Maura Verling at Haulbowline and to successive secretaries in the Dublin Office for dealing so patiently, over a long period, with my untidy notes and correspondence.

Lastly I offer my acknowledgments to the Chairman, Kevin McCourt, and his colleagues on the Board of ISH for giving me the opportunity of paying tribute through this medium to a place and people where and among whom I gained much satisfaction and happiness over many years. To the Board I must, in all fairness, also express thanks for their forbearance in accepting the slow pace at which their chosen historian completed his journey.

Sarsfield Hogan,
March 1979

CHAPTER I

Why No Irish Steel?

The first Irish steel mill was opened some forty years ago in August 1939 at Haulbowline Co. Cork and nearly a century after the closure of the Belfast works which had been using native ore and coal in the production of iron plates for ship building. The venture at Haulbowline can thus be seen as a much overdue initiative in the manufacture of a commodity so vital to a developing economy. The delay in setting up a steel industry in Ireland cannot fail to provoke comment and question, and is all the more remarkable when related to the earlier Irish traditions and achievements in the production and working of iron.

The historical panorama of these Irish activities which preceded the manufacture of steel is largely hidden. Yet when the curtain is occasionally drawn aside, there is plenty to stir the imagination. The medieval delvers of easily accessible ore and the charcoal burners were producing iron which could be shaped into crude implements or weapons. English settlers attempted in the late sixteenth and seventeenth centuries to promote iron-working on a commercial scale during the relatively peaceful interludes between the Irish wars. A notable instance was the Earl of Cork's great iron works at Tallow, from which, according to Gerald Boate, 'he hath profited by above one hundred thousand pounds'. Thousands of acres of oak and giant arbutus around Kenmare were felled to furnish charcoal for the smelters (using imported English ore), which the colony, originally settled there by Petty, operated until the Irish drove them and their skills out in 1689. Wooden sailing ships carried abroad iron from Ireland to the Continent and even to the distant Canaries. There were works at Enniscorthy and elsewhere that were said by

Petty to use more than ten large furnaces and twenty forges and to employ over two thousand men and women until local supplies of timber became exhausted. For many years bog iron ore was exported through Waterford or Youghal to South Wales for the iron works at Merthyr and Dowlais. Much later there were the unknown craftsmen of Creevelea who smelted the local ore with local coking coal and wrought structural iron (some of which found its way into the building of Wellington Bridge more commonly called the 'metal bridge' across the Liffey at Dublin); and later still there were Richard Turner's busy foundries, one at Ballsbridge, where Hume House now stands and another at Oxmanton which supplied a quantity of the iron frame for the Crystal Palace erected in 1851 in London for the International Exhibition. To Turner's craft the TCD tall railings on College Street still bear witness.

These are only a few of the vignettes which caught the eye from somewhat random reading including the works of those who have written on the industries and resources of Ireland during the period from Elizabethan times down to the middle of the nineteenth century, such as the Cromwellian Boate, Sir William Petty, Sir Robert Kane and his notable successors in the same field, Sullivan and Hull. Pre-eminently there are the well-known economic histories of the period by Professor George O'Brien, while more recently (1976) the *New History of Ireland, vol. 3,* by Professor T. W. Moody and others sheds further light. L. M. Cullen's *Economic History of Ireland since 1660* (1972) contains some references to iron-making. Lord Macaulay's account of Petty's experiment at Kenmare should be read, if only for his 'purple passage' describing the beauties of Iveragh *(History of England,* vol. I, chapter 3).

Some of these authors, notably Kane and O'Brien, when dealing with early iron production in Ireland quote freely from Doctor Gerald Boate. He was an apothecary and scientist who came to Ireland as an officer in Cromwell's army. In 1652 he wrote *Natural History of Ireland* which included a description of the location, nature, use and exports of Irish iron resources at that time. Boate fascinates by his detail, particularly as regards the elusive bog ore found in 'low and boggie places, raised with very little charge and very rich of metal'. There is one extract from his book (not quoted by the writers mentioned) which may be summarised here as showing what was the reality of Irish iron-smelting in the mid-seventeenth century:

> The huge furnace being once kindled is never suffered to go out but is continually kept a burning from one end of the year to the other, blown by two pairs of bellows kept in action by a great wheel driven by a little brook or water course. . . . Then there is a lesser sort of iron works, called Bloomeries (mainly in Ulster) where there is a large hearth with similar continuous firing and bellows. . . . There are also forges for casting the iron into cannon, railings and such like as is done by Christopher Wandsworth at Carlow. . . . These 'Hammer Works' or forges where the crude iron is reheated and hammered into the required shapes also have water wheels to drive the hammers. . . . One ton of sowe [pig] iron could be got out of three tons of ore. The hammer works produced a ton of merchant iron from one and a half tons of sowe iron. (Boate, c. 17)

It is interesting to observe the use of water power in the iron-works, presumably the English practice. The continual operation of the furnace required vast quantities of timber to produce charcoal. As Boate remarked 'it is incredible what a quantity is consumed by one iron-work in a year'.

Moody suggests that the improvident use of timber supplies to produce cheap charcoal was one of the reasons for the decline of Irish iron production after its early commercial successes.

Thus there was a long tradition of manufacturing iron at various places in Ireland until diminished natural resources and other economic pressures, including the power of the British coal and iron industries and the lack of local investment, extinguished iron smelting, with the Leitrim and Ulster operations apparently taking longest to expire. The manufacture of iron from local ore was replaced by iron foundries using as their main feed-stock local scrap iron and pig iron usually imported from Britain. There was scarcely a town of any size which had not its own foundry. The Dublin directory of 1850 mentions twenty-nine foundries in that city, including two substantial iron-works on the Dodder banks near Clonskeagh one of which (Messrs Rochford) produced iron keels for vessels up to ninety feet in length. That was shortly after Robert Kane had published his remarkable book on the *Industrial Resources of Ireland* (1844), written some dozen years before Bessemer discovered the means of refining molten iron into steel in large converters.

Bessemer's invention launched a second-phase industrial revolution which, from about 1870, saw steel replace iron for a wide and growing variety of structural and industrial purposes. That process was one in which Britain for a while retained the dominance it had gained in heavy industry since, a century earlier, Abraham Darby had perfected the means of using coking coal to reduce ore to molten iron in big blast furnaces thus ending dependence on the diminishing supplies of wood charcoal. This dominance was intensified after 1783 when Henry Cort invented a method of producing malleable iron and shaping it with grooved rolls. Steel had been a well-known metal for hundreds of years. Until the Bessemer system enabled mass production to be undertaken, outputs of steel had, however, been limited to the quantities produced from refining iron in small crucible furnaces, mainly for sharp blades and other precision tools and implements. Steel-making in that manner as a refinement of iron came from the Eastern world to the West. The famed Damascus and Toledo blades were the precursors of Sheffield steel.

As for this country, there may have been some enterprising Irish iron-founders who undertook steel-making in crucible furnaces, but if so, the record of their activities has escaped this admittedly haphazard investigator. There is, however, a reference by C. H. Kinahan (in RDS *Proceedings*, vol. 3), recording that in 1609, according to Chichester, 'Smiths in Ulster were at work making steel out of native iron which they wrought much more easily than it could be made in England'. Possibly the activity was the same as that recounted in a report dated 24 September 1609 from Sir Thomas Phillips to Lord Salisbury (State Papers, Ireland, vol. 227): 'At Toome [Co. Tyrone] we caused some of the ore to be sent for, of which we caused a smith to make iron before our faces and of the iron he made steel in less than one hour. Mr Broad [a London agent] who has skill in such things says that this poor smith has better pleased him than Germans or others who presumed much of their skill.' One can only wonder about the refining process used in Ulster at that early date if, indeed, the end product of the 'poor smith's' labours was really steel and not wrought iron.

In April 1850 Thomas Jevons was granted an Irish patent, enrolled in the Rolls Office, Dublin, in respect of 'certain improved machinery to be used in manufacturing bars or wrought iron into shoes for horses and also into shapes for other purposes'. This was possibily merely intended as protection for a patent already granted in England. It is, however, some collateral evidence that Irish foundries were by no means industrially unprogressive. Galvanising of iron sections was practised, for instance, by Kerrill at Townsend Street Dublin (*Irish Builder,* April 1887). Edmondson, another Dublin firm, which worked both iron and copper, was awarded the contract for an extensive gas plant at St Petersburg in Russia (*Dublin Builder,* June 1873).

The question in the heading to this chapter is, therefore, not altogether rhetorical. It amounts to this: why should it have happened that Ireland with its long, if spasmodic, experience of local iron production and its subsequent apparently prosperous system of iron foundries based on imported materials, failed to produce a single ton of crude steel or a single ton of steel rolled from imported ingots or billets until nearly a century after Bessemer? – Or, to put it more fairly, until some seventy years after the Siemens basic open-heart furnaces, working on a cold charge of metal, came into use for, inter alia, recycling scrap steel to provide new merchant steel for builders and manufacturers?

Arthur Young, that persistent and enquiring traveller, gave one answer in the context of late eighteenth-century economics. In 1785 he saw the iron-works at Enniscorthy and Mountrath and he compared their state with that of the English iron producers. The Irish works were smelting English ore but suffered greatly from lack of charcoal due to deforestation. They did not work for more than a few months in every third or fourth year. Young concluded that capital would not flow from England nor would the English iron-masters move from the coal fields. 'No manufacturers', he said, 'ever moved from a rich to a poor country nor would the work force' (O'Brien, *Economic History*).

Robert Kane, in *Industrial Resources of Ireland,* chapter 4, refused to accept the proposition that Ireland should remain entirely dependent on its powerful neighbour for the products of coal mining and heavy industry. Anticipating the modern disposition to make consumption of steel one index of a community's economic progress, Kane eloquently described the importance of iron to the advancement of civilisation and wealth. 'Its use has become at once the cause and the measure of industrial progress.' He was not content to assert general propositions as to the extent and availability of Irish ore and coal deposits. He sought to establish detailed costings for the production of iron at Arigna compared with similar costs in England and on the Continent at that time, dealing also in detail with the possibilities of using turf charcoal as the heating medium. Kane emphasised the importance of fuel costs in attempting to establish competitive Irish industries. In the first sentence of his masterpiece he pointed out that in considering every industrial enterprise, the first question arising was 'the amount and cost of the fuel necessary to be consumed in executing the work'. (As to the relevance of Kane's propositions to the circumstances of today, see M. J. Cranley's *Kane Lecture* delivered on 19 May 1976 in University College Dublin.) In 1864 Kane published a paper which included costings to show that Irish industries could be made competitive on the basis of using imported English coal as fuel. When Irish steel was first produced at Haulbowline, it was Yorkshire coal that was used in the gas producer plant which gave heat to the furnaces.

Kane's book was indeed a beacon in Irish economic history. It must be noted, however, that it was written long before the development of steel-making technology. While the arguments for industrialisation based on native resources remained, in general, valid, they lost force in relation to the modern possibilities of viable steel production from importation, wholly or mainly, of the primary materials.

Apart from the economic aspects there was another strand in the argument as to 'why no Irish steel'. After the Union of 1800, British government and commercial interests made a practice of discouraging, either by overt restraints or secret measures, the development of industrial activities in Ireland that might threaten the market supremacy of British producers, including the coal and iron industries. These 'enemies of Irish prosperity' and their misdeeds, were portrayed in the writings of Professor O'Brien on the economic history of Ireland (including his penetrating Introduction to O'Riordan's *Modern Irish Trade and Industry,* 1920). It is true that O'Brien's examples do not refer at all to the decline of iron production as an instance of such a political process. His reference to the ten per cent duty on Irish iron entering Britain (Act of Union 1800) is inconclusive. The countervailing duty against British wrought iron may have given a net benefit to the Irish foundries while these duties lasted.

No evidence of any kind was found of government influence to prevent investment in the commercial exploitation of what remained of Irish resources of coal and iron ore. Equally, one cannot find evidence of any inclination among Irish industrialists to devote effort and investments to the production of steel when technology opened the way for such an enterprise in places where the economic advantages of adequate native basic materials were absent. Comparisons in this field are misleading. Yet the question remains, why should it have been possible for Holland and Denmark to establish steel-making as an essential support for their other main industries, while in Ireland not even the concept, let alone the attempt, appears to have received any serious consideration?

In western Europe both Denmark and Holland came late to the production of iron and steel. Neither possesses ferrous minerals although in Holland there is abundant coal. Denmark is the more apposite example for Ireland, being a small country, primarily an agricultural economy and without iron ore or coal resources. There an efficient iron steel industry was established on the basis of imported materials from its neighbours Sweden and Germany. Danish steel production supports its manufacturing industries, notably ship building. In Ireland, there were demonstrable difficulties both political and economic, which hampered industrial progress. Yet the case of Denmark does seem to prove the point that the absence of native raw materials cannot be claimed as a compelling reason for not having attempted steel production, even on a limited scale, any earlier than the middle of the twentieth century.

There was undoubtedly one Irish industry which developed strongly in Belfast and retained its pre-eminence as iron replaced wood and later when steel plates became essential in the building of ships. As the Harland and Wolff yard grew in stature and financial strength, a question must often have been raised in the board room among men who were leaders in their industry and well aware of the profits to be gained from vertical integration. They imported all their steel requirements from across the water. Would it have been a good idea to instal their own steel-making plant beside the Lagan; not necessarily a full-blown integrated works from ore to finished product but perhaps a modern plate-rolling mill fed by imported semi-finished steel? If such a scheme was never in mind, the problem was clearly recognised. Harland and Wolff sought to solve it, in a financial sense, by gaining control of Colville's, the Clydeside steel firm which was the main supplier of plates to the ship-yard. Lord Pirrie, the shipyard chairman, who entered into marriage with Colville's in 1919, cannot have found the union rewarding. It was dissolved seven years later in the course of efforts to reconstruct the troubled British steel industry (Vaizey, *History of British Steel).*

The suggested interest of Harland and Wolff in producing steel in Ireland is stretching surmise, although the thought is given legitimacy by Hull's plea for an Ulster steel industry as long ago as 1885. There is no evidence that if such a project had been in mind it would have been frustrated by pressures from the British steel producers, reinforced perhaps by government influence. Any plant supported by such large buyers as the Belfast shipyards, between them employing up to 30,000 people in 1919, and no doubt also by the smaller yards then existing in Derry, Dublin and Cork would, however, seem to have been worth while in economic terms, assuming that material and labour costs made competitive sense. As to local materials for steel making (as distinct from the narrower objectives of rolling plates from imported semis) it is proper to note that at least some quantities (no judgment is offered as regards suitable quality) did exist at that time. The official output of iron ore, mainly if not exclusively from Antrim workings, had declined from 150,000 tons annually thirty years earlier but this may have been for economic reasons and not altogether due to poverty of deposits. It has never been seriously argued that anyone contemplating a steel industry in Ireland whether in 1919 or afterwards, would have to do so solely on the basis of native raw materials. Where these existed and were of suitable quality, however, they might have been used with beneficial social and economic results. Apart from coal and ore, and iron and steel scrap, there were ample quantities of limestone for use as a refining agent in steel making. Manganese ore had been mined at Glandore in Co. Cork but this is doubtless an irrelevancy as the deposits were presumably worked out before 1919.

There may seem to be no great point in pursuing further the possibility that the failure to commence steel manufacture in Ireland was due to a sinister influence from Britain directed against the creation of strategic industries in a country liable to be a problem in time of war. The idea is, however, encouraged by other beliefs, widely held at the time under consideration, that the British government placed obstacles in the way of Cunard adopting Cobh (then Queenstown) as a port of call, and also hampered Fords when in 1916 they first mooted their project for a Cork factory. Such British policies were a constant theme with Arthur Griffith, and the following instance is mentioned because Kynochs explosives factory at Arklow, or rather its site, appears again in the story of Irish steel. In the issue of *Nationality* for 31 July 1915, the leading article signed by Arthur Griffith dealt with protests by Irish firms that they were not getting orders for munitions. In this context, Griffith recalled three interviews solicited in July 1907 by Arthur Chamberlain, the chairman of Kynochs who was a brother of the noted statesman, Joseph Chamberlain. Arthur Chamberlain had revealed 'the secret attitude of the British government to any scheme to seriously develop Ireland industrially'. In particular, he had told of an interview with Herbert Gladstone, a member of the Liberal government in power at the time, who had informed Chamberlain that, if the Arklow factory continued, the government would see to it that the factory got as little work as possible. On the other hand, the fullest support was offered to a branch of Kynochs being planned for South Africa. In *Nationality* of 23 February 1918, there is another leader by Griffith: 'In the last few days, England has shown her hand in Arklow. In that town there is a factory of Kynochs which the English government sought to close down in 1907 and now seeks to close down again in 1918'. Griffith's theme was that until Ireland had control of her own Customs, there would be no hope of industrial expansion.

This narrator of the events of those early days turns from economic historians and polemicists and from the blame attributed to England for Ireland's slow industrial development. He searches for evidence (subsequent to Kane's vision of 1844) of any Irish blueprint for future prosperity in which the necessity or even the merit of steel manufacture in Ireland was recognised. Secondary industries (which, in part at least, must rely on imported steel products) were no doubt advocated by the Home Rule party when it became a strong political force from 1880 onwards, as indeed was the exploitation of mineral resources. The quest for economic self-sufficiency was an integral part of the gospel of Sinn Féin when it advanced to influence and power after 1916. Yet one looks in vain in national policies and pronouncements for even slanted references to the possibility of producing steel in Ireland as an ingredient in the desired industrial progress. The need for steel may have been proclaimed but, if it was, that voice cannot now be heard amid the louder calls at that time for other developments and innovations in both agriculture and industry.

It would be unjust, however, to ignore some post-Kane advocates of greater attention to Irish mineral resources who, significantly, referred to iron production as one of the possibilities in a programme of industrialisation. W. K. Sullivan, the distinguished scientist, a disciple of Sir Robert Kane, whom he succeeded as president of Queen's College Cork in 1873, gave evidence on that theme before the House of Commons Committee on Irish Industries in 1885–86, when, as well, he pressed strongly for state aid to technical education as a necessary base for industrial progress. Kane and Edward Hull also gave evidence before that committee. The latter thought there were sufficient deposits of iron and coal in Ulster to supply the local demand for steel, even for the great manufacturers including the shipyards of Belfast. He maintained that steel suitable for ship building could be produced from Ulster ore. (Some further information about that parliamentary enquiry is given in a Note to this chapter.) At an earlier period, in his guide to the Cork Exhibition of 1852, John F. Maguire showed enthusiasm about Irish iron resources. He quoted from the *Belfast Newsletter,* of February in that year, concerning the activities of a Scottish company at Creevelea near Manorhamilton where 'coal and ironstone were being brought to the furnaces for about three shillings a ton . . . it only requires that the district be opened by railway communications to make Leitrim the Staffordshire of Ireland'. (See also Appendix 2, J. H. Murphy's *Ireland,* 1870, concerning iron-works at Arigna.)

In 1902, the Department of Agriculture and Technical Instruction (a recent addition to the British administrative structure in Ireland) published a very useful survey, *Ireland – Industrial and Agricultural.* There is little in the survey about iron-working. Bog iron ore was being exported – 4,321 tons in 1899 – from Donegal and from Mountrath, apparently for use in gas purification plants. In the year 1900 as much as 100,000 tons of iron ore had been raised from deposits at Glenarm and Broughshane in Antrim, all no doubt exported to Britain.

The silence about steel, as distinct from iron, persisted even when the great political changes took place in Ireland following the general election of December 1918. The first Dáil Éireann met on 21 January 1919, re-affirmed the sovereign independence of the Republic and adopted a Democratic Programme which included the following: 'It shall be the duty of the Republic to promote the development of the Nation's resources . . . to exploit its mineral deposits.' In order to forward that industrial policy the Dáil subsequently passed (17 September 1920) what was termed 'a Bill to establish an Economic Council'. There was little such an institution could achieve without financial resources or, indeed, executive powers during the war years. Earlier, in June 1919, the published proceedings of the Dáil recorded that a number of prominent people had agreed to serve on a commission of inquiry into natural resources. That body was active during the years 1920–1922 and published several reports including those on coal resources and on peat. In the report on coal, iron-working was mentioned incidentally. The possibility of steel production in Ireland was not at issue. One of the witnesses, P. J. McAndrew, is recorded as defending the quality of Arigna and Tyrone coal, claiming it was suitable for iron production and testifying to deposits of hematite iron in Antrim, Down, Derry and near Rathdrum in Wicklow. Another witness, E. Neill Clarke, the County Surveyor for Leitrim, recalled the extent of previous iron-working in that area. The word 'steel' was mentioned once in the course of evidence by H. Sweet from Antrim: that 'charcoal' produced from peat could be used in iron and steel manufacture; however, he proposed that the charcoal should be exported to British industry.

At a meeting in June 1920, some impatience was expressed in the Dáil because the Industrial Commission had not reported. The Minister for Finance, Michael Collins, is recorded in the Dáil minutes as urging the Commission to get down to the examination of mineral deposits and as saying 'areas containing valuable minerals should be taken over by County Councils or other public authorities, the idea being that they would become the property of the State and could not be exploited by private syndicates'. Arthur Griffith, Acting President, was inclined to pooh-pooh that idea. 'The Dáil', he said, 'had no resources to undertake the acquisition or working of minerals.' It was for others to provide the capital to promote such ventures. Griffith pointed out that neither the Dáil nor its commission of inquiry had the means, in existing circumstances, of carrying out a full geological survey of the country, something which, indeed, had never been done. A contrast can be seen in that interchange between the seeming radicalism of the youthful Collins and the conservative tendencies of the older Griffith. Within two years both were dead. They never had the opportunity in an independent government to advance their convictions on social and economic questions. (There is a Note on some of Collins's relevant views at the end of this chaper.)

In August 1921, Ernest Blythe tabled a report before the Dáil on behalf of his Trade and Industries Department concerning projects for new industries: '. . . many schemes were submitted to the Trade Department which we felt bound to discourage. These included, for instance, certain mining projects.' No reasons were given nor was steel mentioned. Admittedly excerpts from early Dáil minutes are somewhat remote from the subject of this chapter. The temptation to include them was difficult to resist, those illustrious voices echoing curiously in the debates of today about the respective roles of the state and private enterprise in the exploration, development and exploitation of mineral resources. One point in Blythe's report of 1921 has, however, a little relevance. A sugar factory was proposed for Arklow, using the land and buildings of the former Kynochs factory. A French expert was invited to examine the project and concluded that 'Arklow is about the most unfavourable location that could be chosen.' Sixteen years later Arklow was proposed and discarded as a possible home for Irish Steel.

The final phase of this intrusion into some curiosities of Irish economic history concerns the period between 1922 and 1937, when a project for manufacturing steel was first advanced. During these fifteen years one can discern no evidence that the question of producing even part of the country's steel requirements locally received any attention by government or by private enterprise or was aired by official advisers or others in suggestions for extending the range of native industries. Official trade statistics continued to record imports of steel either fully fabricated and ready for use or partly fabricated whether as components to be finished in light engineering workshops or for use by manufacturers of agricultural machinery and implements or in the building and construction industries. These same statistics also continued to show the volume and value of the steel scrap which was being exported mainly for consumption in foreign steel works although sometimes in re-usable form as rails, pitprops and so forth. It could easily be seen that the volume of exported scrap was less than the volume of steel products coming in and that, in other respects, Ireland apparently possessed few natural advantages for steel making. Self-sufficiency in all types and qualities of steel was clearly out of the question. Moreover, even limited steel making involved great employment of capital and where were the entrepreneurs willing to provide it, even with a promise of government support by tariffs or otherwise?

Be that as it may, the first Irish Free State government under W. T. Cosgrave (1923–32), which in the midst of its many difficulties had to its credit the great Shannon hydro-electric sheme and many potent measures in support of agriculture, was not one to give priority to the establishment of new industries behind protective barriers or to the investment of state money in enterprises which should have been financed by private risk capital. One may be fairly certain that no interests, whether domestic or foreign, came forward to propose steel-making. Nor did they when in 1932 de Valera took office with a definite policy, administered by Sean Lemass as Minister for Industry and Commerce, to promote new private industries behind tariff barriers and to finance public development activities with state capital. The latter included peat production on a national scale, the exploration and development of mineral resources and the creation of new transport undertakings both land and air. As the experiences of 1939–45 proved, it was unfortunate that no proposal for steel manufacture emerged earlier than 1937 when it was too late for the installations to be completed before war broke out and the Emergency was declared.

It is well to be clear on one aspect. Notwithstanding the early Irish experience of iron-working and the optimistic assumptions by Kane and others that abundant supplies of suitable ore, ironstone and coal existed, it is certain that any scheme for viable steel manufacture must rely on importation of many materials and on the adoption of the most modern steel-making and rolling practices to ensure acceptable costings. As to the native resources of appropriate minerals, Bishopp, formerly of the Irish Geological Survey, published a *Survey of Irish Mineral Resources* (1943). Following re-examination only Avoca and Arigna were found to have significant iron deposits. The ores at Arigna, Bishopp said, suffered from high phosphorus, while low concentration meant that a great tonnage of barren rock would have to be mined for every ton of smelting ore produced. The ore deposits at Avoca (Ballard and Ballonaneen) had never been fully developed, but their extent appeared limited. There were deposits of low-grade ore and also of bog ore in various other places in Ireland but only in quite exceptional circumstances would it be practicable to use these as a further source of iron, and then only in small quantities and at high cost. The Antrim deposits already noted earlier in this chapter were, according to Bishopp, more promising. In 1880, 228,000 tons had been mined. The later production on a much lower scale averaged 38.8 per cent iron. As he pointed out in 1943, the discovery of vast supplies of iron ore in other parts of the world made the economic use of Irish deposits, so far as they existed, very doubtful. It was suggested that further geophysical experiments should be undertaken to establish the extent of Irish iron deposits.

Clearly Bishopp's findings, published after Irish Steel Ltd. had begun its short life, provide one answer to the question as to why the possibility of steel-making in Ireland had been ignored for so long. It was not the only answer. There were other factors – policy, economics, priorities, disbelief in the likelihood of profit, failure to recognise both the importance of steel production in any programme of industrialisation and its feasibility without natural advantages. These factors seem to have conspired to bring acceptance of the dominance of British and other foreign steel producers as a permanent fact of life in the Irish economy. All the more credit then to the Dublin ironfounder and scrap merchant who eventually decided to make the attempt which others might, with greater advantage have made some decades earlier.

Note on the parliamentary inquiry of 1885 (see p. 10)

The origins of the parliamentary inquiry of 1885 can be traced back to the inclusion of 'promotion of Irish industries' among the objects of the Irish National League founded by Parnell in October 1882. One of the methods of industrial revival urged by him was legislative protection of Irish industries against British competition.

A House of Commons resolution of 10 March 1885 set up the select committee 'to inquire into the natural resources and the present condition of manufacturing and productive industries in Ireland and to consider by what means these natural resources may be more fully developed and how these industries may be encouraged and extended'. Parnell was a member of the select committee but he appears to have attended only two of its thirty-five meetings. More assiduous in attendance and examination of witnesses was another Nationalist MP, Sexton, the member for Sligo and later West Belfast.

The outcome of the inquiry was far from momentous. A

report of five lines stated that the committee 'had not sufficient evidence to enable them to make specific recommendations'.

Hull, Director of the Irish Geological Survey, stated in his evidence before the Select Committee that there were 185 million tons of workable iron ore in Co. Antrim alone, but that the Tyrone deposits of coal suitable for iron making were less extensive. He was more enthusiastic than Sullivan about industrial development based on minerals including iron ore and coal to support an Ulster steel works. Sir Robert Kane in his evidence sounds somewhat disappointed and disillusioned forty years after his famous book was written. His views in reply to questions about possible state action to promote Irish industries were extremely conservative. Clearly he remained a believer in private enterprise and private capital as the sound basis for industrial development. As regards the possibilities for a steel industry, Kane agreed with Hull and rejected Sullivan but not emphatically. No doubt, being far from young, he grew tired of the probing; he was in the witness chair for three days and answered over one thousand questions.

Note on Michael Collins (see p. 12)

O'Hegarty recounts in *Ireland under the Union* that Michael Collins strongly opposed and indeed endeavoured to suppress the Democratic Programme prior to its adoption by the First Dáil in January 1919, and that he took this stance because he thought that the programme was too radical and socialistic. His advocacy of public ownership and exploitation of minerals in the Dáil in June 1920 seems to be inconsistent with that stance, which he repeated in general terms in *The Path to Freedom* published in February 1922. In that book, while urging an industrial policy which would rest 'on co-operative lines rather than on the old commercial capitalistic lines', at the same time he added 'we shall safely avoid state socialism, which has nothing to commend it in a country like Ireland and in any case is monopoly of another kind'.

Collins, however, in *The Path to Freedom* expressed the same views as earlier about mineral resources. 'The development of mines and minerals will be on national lines and under national direction,' he said, and he appeared to extend this thesis to other natural resources e.g. water power, '. . . so that the profits from all these national enterprises will belong to the nation for the advantage of the whole nation'. Although his book gives a list of industries which could be promoted or supported by state action, steel manufacture was not among them.

CHAPTER II

The First Step

David Frame, with whose name the creation of a steel industry in Ireland is linked, was born in Glasgow of Scottish parents. There his father, John Frame, carried on business as a collector of scrap metal and successfully sold iron and steel scrap to the many foundries and steelworks on Clydeside. David left school early and by 1891, being fifteen years of age, he was already working with his father. He never acquired any formal technical qualifications but he learned every aspect of the scrap trade in both ferrous and non-ferrous metals. Although afterwards he engaged in many business pursuits, the scrap trade from which he sprang was to remain for him an abiding and intense personal interest right up to the end of his life.

In 1892 his father sent David to Dublin to assist in the dismantling of premises which had been bought for their scrap metal content, and also to arrange for the export of this to Glasgow. David saw the possibilities in Ireland and, presumably with parental agreement, remained in Dublin where, in due course, he established under his father's name a scrap-yard at Talbot Place not far from the quay-side. There he dealt in all types of scrap metal and gradually obtained a leading position in the local market and in the export of steel-making scrap to both Glasgow and South Wales. In 1902, Frame took a decision which illustrates his early entrepreneurial flair. Strong's iron foundry in Hammond Lane near the Dublin Four Courts had failed and the property, when offered for sale, seemed of value only to the scrap heap. Frame, having bought the buildings and equipment for the proverbial song, decided to become an iron founder himself. Eight years later, in 1910, he transferred all his business, except the old scrap-yard at Talbot Place, to the extensive premises of an abandoned whiskey distillery with a wharf frontage on the Grand Canal Docks at Ringsend. Needless to say he had acquired this derelict property with the same keen eye for a bargain as he had shown when buying Strong's foundry. Frame brought with him the name Hammond Lane which still remains the label of the substantial manufacturing and trading group in whose activities members of the Frame family are involved to the present day. The early operations at Talbot Place and at the start of the foundry had been on a personal unincorporated basis. In 1904 Frame formed a private limited company which in due course become the public company known as Hammond Lane Foundry Company Ltd. It is the latter company which was used as the vehicle for Frame's interest and investment in the steel industry.

During the Great War the Hammond Lane business prospered. The British steel industry was hungry for scrap and there was a high demand for foundry products leading to diversification of the firm's manufacturing activities. Frame established a dominant situation in the Irish scrap trade which he was to maintain until his death in 1946. His son has said that he was accustomed to enunciate the Carnegie philosophy: 'no one ever went broke by taking a profit'.

Hammond Lane emerged relatively unscathed from the great depression of the early 1930s and by 1935 had established a new activity at Ringsend. This was the production of steel drums for the transport of tar and allied substances by local authorities and contractors. These drums were fabricated from sheet steel imported from Belgium by Frame and as many as 1,600 drums a day were produced for the home market. Keen as always to increase his own margins, Frame observed that much of the profits from the production of the drums went to the suppliers of the Belgian steel and to the shipping company which carried the sheets to Dublin. His first proposal for correcting this situation, which he made known to the Department of Industry and Commerce in January 1937, was the erection of a mill at Bray (where he lived) or Wicklow or Arklow which would roll steel sheets from imported steel slabs. This was to be a small operation, some 10,000 tons of steel a year with a capital investment of

£70,000. Clearly the scheme retained the disadvantage that the raw material had to be imported and that foreign steel producers and shippers might, between them take the cream of the profit.

As a result of discussions with his Belgian agents and with contacts of Hammond Lane in the Welsh steel industry and more especially because of information that suitable plant was available for purchase in Belgium at a steel works which had recently closed, the project soon took a more ambitious shape. During the early months of 1937 Frame busied himself in putting the plan together with the aid of consultants. There were discussions with R. C. Ferguson of the Department of Industry and Commerce, the objective being to get some assurance of government support by protective tariffs and other facilities. Eventually, on 22 May 1937, Frame wrote to the Department putting forward his plan for a new industry comprising the manufacture of a wide range of steel products including flat and corrugated sheets, bars, angles, squares, fencing wire, tees, hoops and so forth. Capital investment of £150,000 was envisaged for an annual production of 20,000 tons. The employment of 400–500 persons at the outset would increase considerably later. Arklow was named as the chosen site.

Although the formal assurance of tariff production did not issue from the Department until August 1937, Frame had clearly received sufficient encouragement from his talks with Ferguson, and possibly from an interview with the Minister, to enable him to proceed with his plans. On behalf of Hammond Lane Foundry Company he entered into purchase agreements with the owners of the plant in Belgium (SA Usines Bonehill, Marchinnes-en-pont, Belgium). The bulk of these transactions was executed in May and June 1937. He also entered into formal agreements for consultancy services and for the dismantling and transport to Ireland of the buildings and equipment. These agreements were also completed in May 1937. At this point Frame had committed his foundry company to the expenditure of some £65,000, of which slightly more than half covered the purchase price of the various items of plant and steel buildings to be re-erected in Ireland.

Even at 1937 values, the prices paid for the items of plant

described in the contracts appear remarkably low and in some instances no more than scrap value, of which Frame was an excellent judge. But in the field of steel-making and the probable efficiency of the plant and ancillary equipment offered for sale, Frame had to rely on the opinions of others. There was no one in Ireland competent to advise him. He was already acquainted with some prominent steel producers in South Wales, who became not only his advisers but his partners in the project for an Irish steel industry. They were Major J. M. Bevan, chairman and managing director of Briton Ferry Steel Company, A. W. Kieft who had been joint assistant managing director of Richard Thomas and Company, and H. P. Lloyd, the managing director of Neath Sheet Steel and Galvanising Company. On the face of it these were experienced men. For specialist consultancy services, Frame and his associates turned to the Loewy Engineering Company of London. The principal of this firm, Ludwig Loewy, a former design engineer with one of the leading German rolling mill manufacturers, was a recent émigré to Britain and had gathered together a group of other German technicians who had fled from the persecution of the Jews. One of the Loewy group, W. Masgeik, subsequently filled a senior post on the technical staff of Irish Steel Ltd.

It was Loewy's task to advise on the purchase of plant, to supervise the installation on an Irish site, to prepare capital estimates and production costings and to establish a basis for commercial viability. In the prospectus for the public issue of shares in June 1938, Ludwig Loewy signed a Certificate that the valuation of the plant and materials which at that date formed part of the steelworks and rolling mills, was £204,884 as compared with £183,000 which Irish Steel Ltd. paid to the Syndicate for these assets. It was scarcely an independent valuation, as Loewy had, in turn, advised all three parties to these transactions, i.e. Hammond Lane, the Syndicate and Irish Steel, a situation rendered even more bizarre by the fact that these three parties were, for all practical purposes, dependent on the vision and energy of one man, David Frame.

As described earlier, Frame appears originally to have conceived a new industry which would be absorbed into the Hammond Lane group of activities. His mind, however, shifted at an early stage to other possibilities, to promotional profits and investment dividends as the objects to be pursued. James J. Davy, a Dublin stockbroker, who was a director of the Hammond Lane company at that time and became a first director of Irish Steel Ltd., has suggested that the financial project was Frame's intention from the outset. The plan was that Hammond Lane would buy the Belgian plant, sell it at a profit to a syndicate to be formed by Frame in which his company would have the largest stake and finally establish a public steel company to which the syndicate would, in turn, sell its assets at a further profit. The steel company would obtain the cash to finance its obligations from an issue of shares and debenture stock for public subscription. Under the scheme Hammond Lane would eventually emerge as a large shareholder in the steel company with the added advantage of being a supplier of scrap to the furnaces and a purchaser of the products of the rolling mills. The chances of worth-while financial benefit for Hammond Lane and for the other members of the syndicate may well have been weaker than those promised by the contemporary estimates and forecasts; such as they were, these chances vanished with the outbreak of war in September 1939.

The private company known as Haulbowline Steel Syndicate Ltd. was incorporated on 17 December 1937 with a share capital of 200 shares of £1 each, which were to be issued at a premium of £999, and with the stated object of acquiring certain assets and contracts of the Hammond Lane Foundry. The subscribers to the Memorandum of Association and also the first directors of the company were David Frame and James Fitzgerald. Of the shareholders shown in the Companies Office file nearly all acquired only one share, each representing a cash payment of £1,000. Hammond Lane Foundry Company got 60 shares issued to them without payment as part of the consideration for transfer of the assets. Other major shareholders were Neath Steel and Galvanising Company (10 shares), Mayer Newman, London associates of David Frame (5 shares), Major Bevan (3 shares) James Fitzgerald (2 shares), McArthur (2 shares) and Kieft (2 shares). Of the two dozen personal shareholders, about half appear to have been Frame and his immediate associates and members of their families, and the rest persons from the Cork area for whose participation Fitzgerald may have been responsible. There was no apparent canvassing for subscriptions. It was a selective process and the chosen few were privileged to be admitted to the circle.

The Syndicate and its membership had no doubt been determined for months before the incorporation of the company which was the legal preliminary to the purchase of Frame's Belgian plant. The agreement between Hammond Lane Foundry Company and the Haulbowline Steel Syndicate is dated 29 December 1937 and is signed by David Frame and his co-director Buchanan on behalf of the vendors and by Frame and Fitzgerald on behalf of the purchasers. The duality of Frame's position is recognised by a clause in the agreement which, inter alia, states that 'the validity of this agreement shall not be impeached on the ground that the company (Hammond Lane) as promoter or otherwise, stands in a fiduciary relation to the Syndicate . . . the company is selling at a profit the amount of which is known to all the members of the Syndicate . . . that the said David Frame should not, nor should the company, be liable to account to the Syndicate for such profit' (Public Record Office File No. 9544). The agreement, in effect, makes it plain that Hammond Lane sold at a profit of £10,000 plant and equipment for which it had paid £35,601 and in addition that the Syndicate undertook to repay to Hammond Lane expenses incurred in respect of transport, maintenance, dismantling, loading etc. of the plant and sums paid to Loewy, Consulting Engineers, under contracts made by Frame in May 1937. The total 'purchase price' was £76,157 of which £16,157 was paid in cash and the remainder by allotment of fully paid shares in the Syndicate.

The first and second schedules of the agreement itemise the plant and equipment sold to the Syndicate and the prices which Frame had paid for each item. The description of the plant is extremely cryptic, but fortunately George Radley, who served in the drawing office of Irish Steel Ltd. and knew the Belgian plant, was willing to furnish his recollection of the buildings and plant as they existed in the original purchase.

There were eight steel-framed and corrugated-sheeted buildings, appropriate to gas producer plant, scrap-yard, furnace bay, casting bay, ingot bay, blooming mill/sheet mill, merchant mill and merchant mill annex, together with seven overhead cranes. All these buildings were erected on arrival at Haulbowline and most of them are still *in situ,* although they have been subject to much repair and adaptation over the years.

The plant and machinery comprised:

1. Four gas producers complete with bunkers, grate drives and mains.
2. Blooming mill with two 650-mm 3– high stands, rolls etc. but without the main drive and motor.
3. Hand sheet mill also without main drive and motor.
4. Three 2– high stands for tin-plate rolling.
5. Sheet bar re-heating furnace and pickling crates.
6. Merchant mill also without main drive and motor.

George Radley, whose engineering skill is matched by his long memory, points out that in 1937 Loewy planned to utilise the useful parts of the Belgian plant and to purchase new equipment for the completion of the general scheme. While a considerable amount of material, particularly for the blooming mill and the sheet mill, was delivered prior to the outbreak of war, no progress at all was made with the open-hearth furnaces except foundations and the construction of the two 50-metre high brick chimneys, so graceful and spectacular a feature of the Cobh sea-scape for many years. Radley says that the merchant mill purchased in Belgium must always have been a 'non-runner' because of the complete absence of layout drawings. As, according to the agreement with the Syndicate, Hammond Lane paid only £1,250 for this mill, the equipment must have been a doubtful asset. It was eventually sold as scrap by Irish Steel Ltd., presumably to Hammond Lane.

The merchant mill had not been given priority in Loewy's plans until the shadows of the impending war lengthened and the further necessity arose of concentrating activity on some section of the programme which could give a quick and certain financial return. Any market survey of steel consumption in Ireland in 1937 or 1938 must have revealed that a much greater demand existed for reinforcing bars and similar sections than for sheet steel for fabrication or for galvanised roofing. Fortunately for the promoters of the steel industry, their conclusion that available financial resources should be concentrated primarily on the merchant bar mill coincided with information that such a mill in Germany was being offered for quick sale by Krupp of Essen. This mill had been manufactured for installation in India (for rolling copper bars, according to Radley) but that project broke down. Hence it was a new mill and Radley comments 'although the billet furnace was coal fired and the mill operations 100 per cent manual, this Krupp mill, when erected at Haulbowline, was considered quite modern at that time'. In May 1938, following inspection by Loewy and with Frame's agreement (he was in South Africa at the time), the merchant bar mill and its ancillary equipment were bought by the Syndicate for about £75,000. This outlay, together with the money spent on buildings, cranes and foundations and, of course, the value of the state leases of land and premises on Haulbowline Island, seems to be the sum total of the effective investment (in the sense of being likely to yield an early cash return) achieved by the original promoters from the capital resources at their disposal. The Krupp mill gave good service although it had serious deficiencies in its dependence on manual operation with consequent low through-put rates and a degree of risk of injury to workers engaged on 'looping' operations.

There is something of a mystery about the choice of Haulbowline as the site for Frame's steel venture. A location in Co. Wicklow was the basis of his first approach to the Department of Industry and Commerce. The written proposals in May 1937 envisaged the site at Arklow, although it is quite clear that as early as January of that year, Frame had approached the Commissioners of Public Works about the possibility of a lease at Haulbowline. The Department of Industry and Commerce were seemingly unaware of this approach and indeed Frame was understood to maintain that 'he would not go to Cork under any circumstances'. Arklow it was to be and R. C. Ferguson noted 'I discussed with the Minister and he said he would like the scheme to proceed and would accept Arklow as the site.' The first recorded reference to Haulbowline appears to be a letter from the Cobh Trade Unions Council to the Department of Industry and Commerce dated 11 May 1937 urging that the proposed steel mill be erected on Haulbowline.

There was every reason why David Frame should have had a preference for Arklow. He personally (not Hammond Lane) had acquired for its scrap value the Kynoch Explosives Factory and the land on which it stood on the estuary shore at Arklow. This factory had gone out of business when the boom years during the Great War came to an end. This extensive property was clearly attractive to Frame as the home of the steel works, but the harbour facilities were so inadequate that a substantial investment in their improvement could not be ignored in estimating the cost of a project at Arklow.

There can be no doubt that Frame and the Syndicate received encouragement from the government to opt for Haulbowline as the site for their enterprise. There had been endemic unemployment in the Cobh community since the British had closed their naval maintenance and engineering services at Haulbowline Dockyard. David Duff Frame, who joined the venture as a member of the Syndicate and a first director of Irish Steel Ltd., has related that de Valera himself, then head of the government, pressed his father to go to Haulbowline. Writing in June 1940 Lemass, then the Minister for Supplies, accepted responsibility – 'it was I who urged the selection of the Haulbowline site'. Eight years later when in opposition and urging his successor, Dan Morrissey, to support the Irish steel industry, Lemass said 'it may be that its location at Haulbowline was a mistake' but he added that the disadvantages would diminish according as outputs increased and as more satisfactory shipping services to the island were established (*Dáil Debates*, vol. 112, col. 1125).

The truth is probably that at a very early stage Frame had his eye on Haulbowline as a possibility and that he allowed himself to be dragged reluctantly in that direction so as to improve his bargaining position with the government. It is not clear when Frame first became associated with James Fitzgerald but it can be surmised that he was a potent influence in favour of Haulbowline. More usually known as Seamus Fitzgerald, this man was a veteran of the Easter Rising and also fought in both the Anglo-Irish War of

1919–21 and the subsequent Civil War. While an officer in the Irish Republican Army, he had been elected to Dáil Éireann by East Cork and was one of those who, in word and deed, opposed the Anglo-Irish Treaty of December 1921. When de Valera's party came to power in 1932, Fitzgerald's political influence increased and, as well, he was prominently involved both in the business life of Cork, where he had an electrical firm, and in the Cobh community where he lived, unceasingly advocating its betterment. As chairman of the local Urban Council in 1924 he sponsored the restoration of Queenstown's ancient Irish name of Cobh. Fitzgerald was, in 1937, chairman of the Cork Harbour Commissioners and of Cobh Urban Council and also a government nominee to the board of the National Air Company. A man of importance and influence, and keenly interested in promoting employment in Cobh and the Cork Harbour area, he appears to have been, apart from David Frame and his Welsh associates, the one most actively involved in the Syndicate and in the formative period of Irish Steel. That he influenced the choice of Haulbowline and that his involvement in the project was an advantage to Frame in his negotiations with the government seem to be beyond question.

The decision to base the project on Haulbowline was, in the circumstances, inevitable although it cannot have been reached without misgivings, at least on the part of Frame's technical and financial advisors. Except for Cork customers the factory was not conveniently placed in relation to its potential market. There were obvious disadvantages in the island situation. Access from the mainland for day to day services and movement of personnel and goods would depend on ferry and barge transport and the vagaries of weather. Efficiency and costs were likely to suffer. On the other hand the existence of ready-made deep-water wharfage under the steel factory's own control, a facility essential for bulk movement of materials inwards and outwards both during the construction stage and in the development of trade, was too valuable an advantage to be missed. A similar harbour elsewhere could not have been provided without capital expenditure far beyond the scope of the promoter's plans. Also, in the vicinity there was a pool of labour hungry for work and some of these workers could be accommodated in numerous small houses on Haulbowline left vacant since the departure of the British.

The decisive element in the choice was that the government and Fitzgerald wanted Haulbowline to be used in order to reduce unemployment in Cobh. The end result was that Frame obtained his lease of state property, his tariff protection and the embargo on scrap exports, not to mention the exclusive new manufacturing licence under the Control of Manufacturers Act which he needed in order to secure his flank against all possible rivals who might be working in the same field.

Of the advantages gained by the promoters, a substantial one was the statutory provision enabling scrap exports to be controlled. The Scrap Iron (Control of Export) Bill was introduced in the Dáil towards the end of 1937. As explained by the Minister for Industry and Commerce on the second stage of the Bill (*Dáil Debates,* vol. 70, col. 261 *et seq.*) the need for the measure arose from 'the danger that our iron founders would be deprived of scrap iron by abnormal exportations on account of the rapid rise in the price of scrap in external markets'. The Minister made it clear that he was seeking a power which might not have to be used. The only dissenting voice raised was that of Deputy James Dillon who objected to the system of control in principle. He alleged that the result would be to force small dealers to sell their scrap to Hammond Lane and Waterford Foundries at any price these firms cared to offer. The Minister had his way, however, and the Bill was passed into law without a division on 16 February 1938.

From reading the debates the curious fact emerges that scrap steel was never mentioned, although that commodity as well as scrap iron was expressly included in the provisions of the Bill. The Minister was then aware of the plans for a steel industry but gave no hint of this to the Dáil. At all events the power to control exports of scrap steel was destined to be the most important, if at times controversial, element in the new legislation.

Except for its relevance to the control of export of scrap there was indeed no reason why parliament should, at that stage, concern itself with the proposals to manufacture steel in Ireland. It was a matter of private enterprise, for which parliamentary approval was not required unless state financial assistance was proposed. Between 1938 (when Irish Steel Ltd. was established) and 1942, no mention of the industry occurs in the parliamentary records except an incidental reference to a new import duty on steel in the Budget of 1939 and three parliamentary questions, one in May 1941 and the other two in March and July 1942 by Deputy Hurley, the member for East Cork, who was concerned about the close-down of the steel mills and the consequent failure to supply horseshoe iron required by the farmers. It is noteworthy that to one of these questions, Sean McEntee, who had temporarily replaced Lemass as Minister for Industry and Commerce, stated in his reply that 'the extent to which the steel works might be a national asset is a matter for consideration'. This was in reply to the Deputy's suggestion that the state had some duty to preserve this 'national asset' and the Minister was patently less than enthusiastic. (*Dáil Debates,* vol. 83, col. 819).

This chapter began with David Frame and it may be fittingly concluded with some further words about the man who made the first effort to establish a steel industry in Ireland. He did so towards the close of a long and successful career in a business he had built up step by step through his own abilities and endeavours. In 1937 the man who, forty-five years earlier, had come from Glasgow to Dublin, married an Irish wife and reared a large family, still retained all his former urgency of purpose together with opportunism and a shrewd calculation of advantages and risks. His failure at Haulbowline was due to a combination of circumstances too damaging even for his ingenuity to overcome. Whatever mistakes were made should not be allowed to detract from the recognition due to David Frame for his concept of an Irish steel industry and for his personal efforts to achieve success in face of manifold difficulties. Contrary to the impression of him, among certain critics, as a tough and inflexible operator whose business acumen provoked the usual jealous or carping comment, he was applauded by a far greater number, attracted warm personal friendships and proved himself both generous and public-spirited as shown when he handed over the Kynoch site to Arklow Urban Council and land at Bray Head to the local council, both for amenity development. David Frame died in 1946. Over thirty years later, his name still survives in context of the steel industry. His son David has been since 1967 a director of the state-owned steel undertaking, while his grandson, bearing the same name, is managing director of the company, Hammond Lane Metal Company, which from the outset has been the principal supplier of scrap to the Haulbowline furnaces.

CHAPTER III

Haulbowline

Haulbowline, the home of the Irish steel industry, is a small island, if the term may still be used since the road bridge to the southern mainland was opened in 1966. It is situated in Cork Harbour about fifteen miles down river from the city and almost directly opposite the town of Cobh on the north shore. The distance between Cobh pier and the landing stage on Haulbowline is about 1,000 yards across water which, near the island, includes the main deep channel of the river Lee. At Cobh there is a railway terminal connected with Cork, and also limited wharfage facilities for transferring goods to and from the railhead.

The origins and development of Haulbowline are relevant to the story of steel-making because the works have been erected on the reclaimed ground between the old Admiralty Stores and the western rim of the basin and graving-dock. Considerable technical problems including persistent tidal flooding of the site were experienced both at the time of the original excavations by Irish Steel Ltd. and subsequently and more intensively when deeper foundations were required during the development by ISH in 1960–61.

At the latter reconstruction a much-travelled site engineer, employed by the consultants, complained bitterly that it was the first time he had been asked to build a steel works floating on the sea. His frustrations no doubt made him exaggerate. Similar difficulties and costly solutions were experienced at the same period by the engineering team that constructed the large integrated steel works at Llanwern in South Wales on reclaimed land at the mouth of the Usk.

The original island was about thirty-five acres in area, with a conspicuous hill at the west end, being for the most part an out-crop of limestone rock supplemented by man-made land. It has been extended by reclamation and is now more than twice its original size. Being so strategically placed in the harbour and dominating the river approach to Cork, the island has a long history of use as a fortification and naval base. As a British naval depot, Haulbowline replaced Kinsale, the transfer of that establishment being completed in 1815. The story of the island is well told by Lt. Commander Niall Brunicardi in his illustrated booklet published in 1968 by the Cork Historical Guides Committee. This fascinating piece of historical research covers not only Haulbowline but Spike Island and Rocky Island too, which also formed part of the fortification system.*

Large store houses and a victualling yard and wharf were constructed by the Admiralty on reclaimed ground at the east of Haulbowline during the Napoleonic wars of the early nineteenth century. These large and most solidly built store-houses are still in use and one of the buildings has housed the administrative offices of the steel company since its inception. In 1865 work was begun on the reclamation of a much larger area, to enclose a nine-acre harbour and wharves, together with a graving-dock which was extended in 1911 to a length adequate to receive the largest British battleship. The area of land necessary to enclose the harbour and dock and to provide for wharves and dockyard installations was over thirty-five acres, the fill being supplied by rock excavated from the original island and other material transported from the mainland. It was a long job finally completed in 1887. The labour was mainly supplied by convicts from the prison on the adjacent Spike Island, many of them political prisoners, particularly after the Fenian rising in 1867. The gangs walked to Haulbowline on a temporary wooden trestle foot-bridge and were required to do very hard labour indeed in handling the massive rocks and boulders used in the filling of the site.

The problems in constructing foundations for the steel

*The name Haulbowline has no nautical significance as might be assumed. It is derived from *Halebolin* which itself was a corruption of the original Norse name for the island, *Aalbolig* meaning the haunt of the conger eel (Brunicardi). In the Irish State Papers for the opening years of the seventeenth century, the island is mentioned in a variety of spellings: Haleboling (October 1601); Halebollen (February 1602); Halboling (April 1602); Halleboline (August 1602); Haelboling (May 1603); Halebolinge (September 1606).

works were described in a paper presented by E. M. Murphy, B.E., consultant to Irish Steel Ltd., at a meeting of the Irish Institute of Civil Engineers in March 1940. This paper is summarised in the *Irish Builder and Engineer,* 13 April 1940. Subsequent to the more extensive and complex civil engineering works of 1960–62, P. Herlihy, B.E., managing director of the contractors Ascon Ltd., gave a detailed lecture with slides to the same Institute, which, inter alia, illustrated the exceptional difficulties faced by both planners and contractors in overcoming the site problems while, at the same time, permitting steel production from the existing installations to continue for as long as possible. This lecture was subsequently published by the Institute.

These engineering difficulties for the steel industry were transient and were successfully overcome. A disadvantage of another kind arose from time to time due to the restricted area allocated to the industry on an island and harbour primarily devoted to the needs of the naval establishment whose interests are paramount when matters of defence and security are involved. Some scope for development has been obtained in recent years by the steel company through the purchase of land on the southern mainland, the cession by the Department of Defence of Rocky Island, through which the road bridge passes, and the continuous creation of new land, mainly at the south-east of Haulbowline, slowly but surely based on the dumping of slag from the furnaces.

Haulbowline had been Crown land and was transferred from the United Kingdom to the Irish Free State in 1922–23, when the Commissioners of Public Works in Dublin assumed responsibility. It was these Commissioners whom David Frame approached early in 1937 when seeking a site for the manufacture of steel. Negotiations were protracted and the actual leases not executed until 30 May 1938. On the following day Frame assigned his interest in the leases to Irish Steel Ltd. At the liquidation the leases were included in the assets which, as will be related later, the receiver sold in June 1947 to the new company established under state auspices.

Even by the standards of the time, Frame would seem to have obtained what he wanted on easy financial terms. No doubt he would have preferred to purchase the freehold but that was not possible in the case of state land. He secured the longest term of a lease permitted by the State Property Act, ninety-nine years commencing on 1 May 1938. The principal lease was that relating to about ten acres between the accommodation water-way in front of the old Naval Storehouses (known as the East Camber) and the western rim of the Basin and Dry Dock, roughly the area occupied by the former Admiralty dockyard installations on that side of the Basin. The steel company (ISH) was permitted in 1956 to fill in the East Camber so as to provide additional space. The original rent under the lease was £350 a year for the first five years, £750 for the next five years and a final rent of £1,350 from 1 May 1948. The rent was exclusive of rates, taxes and of the entire maintenance of the property which was to be borne by the lessee.

The main lease included provision allowing the lessee facilities for loading and discharging vessels, subject to an additional annual payment of £60. Irish Steel Ltd. were also given a licence, subject to an annual charge of £15, to take fresh water at will from the naval supply, which was fed by submarine pipes from the Cobh urban system.* There was no independent fresh water supply on Haulbowline and until recently the quantity available for the steel industry, after Department of Defence requirements at Haulbowline and Spike had first been met, was at times embarrassingly low. While sea water, passed through a cleaning and filtering station, has been extensively used for cooling purposes in the steel industry, some cooling operations, notably in rolling mills, required fresh water of necessity, and very large quantities in the case of the rolling of sheet steel. The latter project was delayed and eventually abandoned for reasons which included the fresh water problem.

The dockside facilities available to the steel industry at Haulbowline comprise the full length of the west side of the basin and graving-dock, together with Spencer Jetty which

*The ordnance drawing of the plans for the first fort on Halebolin (*c.* 1602), as shown in *Pacata Hibernia,* indicates 'a well of fresh water' but it no longer exists. The building and garrisoning of a fort on the island was actively opposed by the Mayor and people of Cork, whose consent, as citizens, had not been sought. This opposition was severely dealt with. William Mead of Cork (presumably a ring-leader) was arraigned for high treason but acquitted by the jury. The jurors were then impeached by the King's Council and punished for their contumacy (*State Papers* June 1603).

lies beyond the mouth of the basin. The jetty was rebuilt and extended by ISH in 1962. The rail system for travelling cranes is continuous from the dockside to the jetty. Adjacent to Spencer Wharf there is a separate pier for use by the ferry boats, also constructed in 1962.

In general, small coasters and barges use the basin wharf. Larger vessels are discharged or loaded at the jetty. The depth of water in the basin is twenty feet and at the jetty thirty feet. If necessary, the jetty could, without unreasonable cost, be extended a relatively small distance to the deep channel of the river at a point where the average depth is sixty to seventy feet. The depths given are as at low water of ordinary spring tides. The tidal rise, on average, is nine to twelve feet at neap tides and up to fourteen feet at spring tides.

The remaining original lease of 1938 concerned buildings, also for a term of ninety-nine years. It included thirty-one residences and a number of the large naval stores, subject to an annual rent of £375. Two of these stores (Nos. 10 and 11) became the administrative block of the steel company, but at first two floors of No. 10 were excluded, being in use by the naval services for church and chapel as they had been by the British. When the Garrison Church of Our Lady, Star of the Sea, was opened in 1956, the top floor eventually reverted to the steel company. It is said that Mass was celebrated for many years, probably over a century, in the room where the executive staff and their guests now take meals. One never heard, however, that the long shadow of former piety influenced conduct or conversation around the table.

The buildings lease of 1938 became ineffective almost as soon as it was written. At government request, the steel company returned most of the residences and some of the stores to the Department of Defence, these being required for garrison purposes during the Emergency. Since then, by various agreements and understandings over the years, the use and occupation of the residences and other buildings covered by the original lease have been substantially altered. Only a few of the houses are retained by the steel company and not all of these for residential purposes.

In addition to leases from the Commissioners of Public Works, the steel company (Irish Steel Holdings) has obtained facilities from time to time from the Minister for Transport and Power, whose Department has statutory control over foreshores, for the making and use of new ground created by the dumping of slag. There have also been other minor reclamations to enable the area of existing factory installations to be increased, as when the merchant bar mill was extended during the interim development of 1971–72. The total area for which Transport and Power (after consultation with the Cork Harbour Commissioners) have so far given such permission and foreshore leases is approximately forty acres, of which twelve acres had been reclaimed by the end of 1976.

Being an island, Haulbowline was entirely dependent on water transport. The British Admiralty had no shortage of vessels for movement of personnel and stores. At that time according to Commander Brunicardi, 'regular communication for civilians was provided by the famous Green Boats, paddle steamers of the old Cork, Blackrock and Passage Railway . . . train tickets could even be bought at the "railway station" on Haulbowline'. Nowadays, the naval establishment and their families, and civilian workers employed by the Department of Defence, continue to be transported in naval launches, but the steel industry has to make its own arrangements. Irish Steel Ltd. possessed a small cabin launch and two barges which were among the assets sold by the receiver to the new company. The latter found it convenient to employ a contractor for day and night ferry services and for barging material to Cobh and Cork. These contractors were the firm founded by the late Daniel Fitzpatrick, whose father was works superintendent in the civilian force employed by the British Admiralty. He was one of a family resident on the island for many years but of much more ancient Cobh origins. Of the same island family was Daniel's younger brother, Christopher, who became the first general manager of Irish Steel Holdings.

For industrial purposes transport of personnel by ferry is subject to the hazards of severe weather causing delays. There have been occasions when the day shift has been marooned while the night shift was unable to leave the mainland due to storm or fog. During the great storm of 1 November 1959, the last ferry of the day left Haulbowline at six p.m. for Cobh with two hundred passengers, while the 'Black Ball' of danger flew from the naval masthead at the Signal Station. One and a half hours later, as those on board vividly remember, the boat was still at sea having failed to reach Cobh, or Rushbrooke, or Monkstown, where all the quays were awash. Eventually, it was possible to land up-river at Passage by the expedient of jumping ashore in batches, as the *Ingot,* unable to tie up on account of the storm, was brought alongside the quay under power in a series of skilful manoevres. A combination of high spring tide and south-easterly gusts of quite exceptional velocity was responsible for this somewhat dramatic episode.

It is almost impossible to consider the circumstances of the steel works on Haulbowline without, at the same time, thinking of the town of Cobh. The community among the terraced streets and villas crowned by the great Cathedral provides the main bloodstream of the industry and fills the role of its guardian in a family sense; the ever-open front door of Irish Steel still faces Cobh but, for its bulk materials and its promise of exports, the industry must live by the sea beyond the harbour heads leading to the great shipping lanes of the world.

The Cove of Cork was, for centuries, a small fishing village with few inhabitants, but well known in the annals for arrivals and departures at the safe anchorage beside it, invaders and conquerors arriving and sad exiles departing—whether as 'Wild Geese' to fight in foreign lands or as emigrants fleeing from persecution or famine. It was under the name of Queenstown, given at the time of the royal visit in 1849, that it grew to urban stature with a population largely dependent on the military and naval establishments in Cork Harbour but enjoying at a later date further profit and prestige when the town was included as a regular port of call by passenger liners sailing between Europe and America.

The urban population of Queenstown veered between 10,000 and 12,000 during the period from 1870 to the end of the Great War. During that war the dockyard on Haulbowline, at its peak, employed over 3,000 civilian workers while others found occupation in the many shore establishments maintained by both the British and American navies. When hostilities had ended and when, in a few years, the transfer from the British to the new Irish government occurred, there was a rapid decline in the fortunes of Cobh. The population had fallen to 7,000 by 1926 and in 1931 it was returned as slightly over 5,000. Unemployment in the town and its environs became a constant social problem. The once busy sea-port could be observed as a gaunt and depressing place of vacant and run-down buildings, the shell of its former glory, with scarcely an echo from the 'Holy Ground' where sea-dogs and sailor men had engaged in the revelries and romance that are immortalised in the popular song. It was to this scene that the steel industry came and provided the transfusion which gave renewed life and hope to Cobh.

Long before the steel industry became a possibility the Irish government had been concerned about the use of Haulbowline as a means of reviving employment in Cobh. The island and its buildings and dockyard equipment became the property of the Irish Free State under the terms of the Anglo-Irish Treaty of December 1921 but it was not until March 1923 that the transfer formally took place. The Commissioners of Public Works then assumed responsibility for the property under the political direction of the Minister for Finance. This situation continued until the Department of Defence was given control of Haulbowline after the outbreak of war in September 1939. The *Dáil Debates* of 1924–26 throw some light on the position in the early years. In October 1924, in reply to questions by Deputies Hennessy and Cooper as to foreign interest in the dockyard and the chances of re-employing the many idle men in Cobh, Ernest Blythe, Minister for Finance, said that British and German firms had recently visited Haulbowline but no proposition had been made by them to the government. The Board of Works employed a care and maintenance staff to keep the dockyard installations in good order and this staff was used to clean and repair six Canadian trawlers (possibly anti-submarine 'Q' ships) still lying at Haulbowline since the war, prior to their sale to Spanish and French purchasers. In November 1926 Deputy Hennessy was again concerned about Haulbowline, stating that it was 'widely believed' that the British government had vetoed an offer for the dockyard by a German firm. Blythe replied that 'no offer for the dockyard or any portion of it has been made since 1922 and there is no question of interference by the British government with the letting of the place'. The point of that interchange in the Dáil can be understood by recalling that at the time, and until their Treaty rights were abrogated by the de Valera-Chamberlain Agreement of 1938, the British retained a military garrison on Spike Island. They might conceivably have been concerned if the Haulbowline dockyard, a few hundred yards away, was occupied by foreign operators although, as Blythe said, they had no right to interfere.

Subsequent efforts to promote civilian employment were minimal and apparently transient. Haulbowline Industries Ltd., conducted by the Hill family, were given in 1928 a licence to use the dock and slip-way for ship breaking prior to their transfer of the business to Passage. Cork Dockyard Ltd. had licences to bring small vessels into the dock for repair or refitting until these facilities were terminated by the Department of Defence. Daniel Fitzpatrick, the contractor for the ferry services, had a letting of the southern boat houses and slip-way from which he conducted his repair and transport operations until he moved to Cobh as Marine Transport Services Ltd., operating from Atlantic Quay. In 1931 and 1933 a new company, Cork Harbour Oil Wharfs Ltd., obtained leases of the Haulbowline coal depot and quay together with oil tanks and houses. Under government licence the company proposed to refine crude petroleum for the manufacture of petrol, kerosene, white spirit, gas oil, fuel oil and other products. Considerable capital was expended by the company on additional tanks, equipment and on the construction of a new wharf for berthage of oil tankers. About one hundred men were employed during the years 1936–40. Due to wartime restrictions, the venture failed and the Department of Defence eventually took possession of the installations. The Department has leased some of the tanks to the steel industry for storage of both water and fuel oil.

After 1947, when the state purchased the Haulbowline steel works, Cobh became greatly dependent on that industry for male employment. Although it is a national and not a purely local industry, the enterprise became in a special way a part of the resurgence of Cobh both from the employment afforded and from the support which the steel company and its people have given to community efforts. In recent years other industries have taken root in the lower Harbour area: there have been, for example, the new dockyard at Rushbrooke, the oil refinery at Whitegate, the chemical plant at Shanbally and the fertiliser factory beside Marina Point. These events, however, are likely to modify rather than remove the long-enduring interdependence of Cobh and the steelworks.

CHAPTER IV

The Short Life of Irish Steel Ltd.

Irish Steel was registered as a public company on 5 April 1938. The subscribers to the Memorandum and Articles of Association of the company included leading members of of the Syndicate, namely David Frame, his son D.D. Frame, Seamus Fitzgerald, James Davy, Major Bevan and E. P. McCarron. The last named, who was to be appointed secretary of the company under a five years service contact, was a man of some eminence and much administrative experience, having been the civil service head of the Department of Local Government from 1922 until he retired prematurely a dozen years later, when, it is said, his views and those of his political chief could not be reconciled.

The authorised share capital of Irish Steel Ltd. was £300,000 in shares of £1 each. It also had ample powers to borrow money, if necessary, by the issue of debentures or debenture stock.

It was not without significance that the registered office of the company was at 111 Pearse Street, Dublin, the same address from which the Haulbowline Steel Syndicate conducted its business. It was, of course, also the home and registered address of the Hammond Lane Foundry and the head-quarters of David Frame. When he and his colleagues were replaced by new directors in 1942, the registered office was moved to Haulbowline.

The immediate tasks facing the company were to complete the purchase and installation of the plant recommended by Loewy to Frame and the Syndicate, to raise money to pay off the Syndicate and to provide the further capital required by the industry. Loewy, with the assistance of Belgian and German plant erectors and of his own technical staff, pressed ahead with the work of the assembly and installation of the plant purchased by the Syndicate. The construction of the foundations of the buildings, furnaces and rolling mills was undertaken by Irish contractors under the supervision of Loewy as part of his general consultancy assignment. E. M. Murphy B.E., who had considerable experience of local authority projects, was appointed civil engineering consultant. The civil contracts were placed with McNally and Company, a Dublin firm.

The financial arrangements became crucial as soon as Irish Steel Ltd. had been formed. The support of the Munster and Leinster Bank had already been enlisted for bridging finance and this bank was to play a key role in the subsequent funding. The first directors included the seven subscribers to the Memorandum of Association, i.e. the six already mentioned together with H. P. Lloyd of Wales. There were three others, A.W. Kieft, A. S. Huxley also of Wales, who was the managing director designate and James C. Dowdall, a director of the Munster and Leinster Bank, very prominent in the business life of Cork and formerly a Senator in the Irish parliament. Towards the end of June 1938, the directors issued a prospectus inviting subscriptions to an issue at par of (a) £250,000 5½% First Mortgage Debenture Stock and (b) 200,000 shares of £1 each. The Munster and Leinster Bank were, as trustees for the debenture stock, holders of a floating charge on all the assets of Irish Steel Ltd. The shareholders were, of course, entirely unsecured.

Because of bank underwriting, the issue was fully placed. The bulk of the debenture stock remained with the underwriters, the Munster and Leinster Bank and the Royal Bank of Ireland. Similarly only some 6,000 shares were bought by the public, in fact by seventy-four individuals, the largest subscription being £500 and sixty-four subscriptions being for £100 or less. Of the remaining shares, 125,000 were allotted to the Syndicate, 1,005 to the directors and 67,632 to the banks as underwriters. The banks were paid £9,720 as underwriting commission and for their services as an issuing house.

The Syndicate paid cash at par for their 125,000 shares. This cash was part of the £188,000 which Irish Steel had agreed to pay for the property and benefits transferred by the Syndicate. The shares so purchased were distributed *pro rata* among the members of the Syndicate. The Hammond Lane Foundry Company received the lion's share, a total of 70,740 shares in Irish Steel. The Haulbowline Steel Syndicate Ltd. was wound up later in 1938, any residual cash in the company, after expenses, being shared by the members.

Permission for dealings in the Irish Steel shares and debenture stock was given by the Dublin and Cork stock exchanges. There is no record of any market dealing in the shares nor any day-to-day quotations. James Davy recalls that there were dealings in the debenture stock in the early days before Irish Steel defaulted on its interest payments. The sparse particulars from the Public Record Office File No. 9606 show changes from time to time in the ordinary share holding. Some were due, no doubt, to inheritance or private transfers. Others require a different explanation, as for instance the 8,409 shares shown to be held by the New Ireland Assurance Company in a return for 1945. Unfortunately the share register of Irish Steel has disappeared along with the company's minutes, journals and other records.

The main point in this account of the public issue is that so few people held shares in Irish Steel who were not otherwise involved as promoters or contractors or bankers (with hopes of advantage and profit beyond those available to a person buying shares as an ordinary market investment). When the company failed, the loss to private shareholders, although unfortunate, was not heavy or widespread.

Of the ten directors all except J. C. Dowdall and Harry Ponsonby Lloyd were shown by the prospectus to be personally interested in the promotion and the public issue under contract or agreement. A. W. Kieft, who had much experience in the British steel industry, (see Vaizey, *British Steel,* pp. 61, 78, 80) provided the testimony on which the directors relied as showing the soundness of the project. His letter dated 3 May 1938 was reproduced in the prospectus:

> Dear Mr Frame,
> As requested I have pleasure in giving you my views as to the steel works in Haulbowline. The location of the mills is admirable and the market for products is clearly shown in your official figures. Continuous deep water for berthage

at the wharves is a feature of potential economy in transport of unique value. To my personal knowledge the plant and equipment has been purchased under most advantageous terms under expert guidance. It is my view that when the works are completed the company will possess a well planned steel works and rolling-mills – unrivalled – which will include the most modern features based on practical experience of economical working and high standards of production.

Yours faithfully,
A. W. Kieft
Formerly Joint Assistant Managing Director
of Richard Thomas and Co. Ltd.

No one would now reasonably challenge Kieft's opinion both of the advantage conferred by the extensive and deep wharfage at Haulbowline and also of the size of the then home market for the proposed products. The official import figures for such steel products in the year 1937, as quoted in the prospectus, showed a total quantity of about 84,000 tons at a declared value of nearly £1.4 million. The buoyant opinions as to the future conveyed by the rest of Kieft's letter have now, alas, a very hollow sound as the works planned were never completed. Indeed in May 1938, when his letter was written, the possibility of completion within the estimated investment of £450,000 must have begun to seem at the very best doubtful.

As stated in the prospectus the promoters hoped 'to go into partial production in six months and full production in twelve months'. In the early stages an output of about 50,000 tons per annum was envisaged, to be increased gradually to the plant's (planned) full capacity of 85,000 tons each year. One must not blame the promoters for appearing optimistic; would they have otherwise committed themselves to the venture? Forty years later an appraisal of these events suggests that David Frame and his commercial associates placed reliance on forecasts and estimates supplied by their technical advisers which rapidly became impossible to fulfil.

Twelve months after the public issue, due to a combination of factors both physical and financial, only one production unit of the complete steel works described in the prospectus was ready for operation. This was the Krupp bar mill and auxiliary equipment. Pending the installation of open hearth furnaces to produce steel ingots from scrap and of the blooming mill to roll these ingots into billets (or into sections suitable for rolling in the sheet mill when completed) the Krupp mill could be fed only from imported billets. In due course contracts were placed for the supply of billets from Belgium.

It was in these circumstances that with considerable publicity the new Irish steel works went into production in late August 1939. Sean Lemass, Minister for Industry and Commerce, carried out the official opening on 24 August in the presence of a distinguished company, which included the Bishop of Cork, the Lord Mayor and many public representatives. All the Irish daily newspapers on the following day carried special features heavy with trade advertisements and full descriptions of the plant and of the range and qualities of the steel products, together with photographs. The material inserted by the company's advertising agents emphasised the comprehensive and modern character of the equipment as well as the assurance of valuable markets and the prospects of further expansion. 'The Haulbowline Mills', it was stated, 'are not only the newest in the world, the plant is the most modern in existence.' It was an impressive display of confidence in the future of the industry, but the plant so described in the article by H. F. Brennan of the advertising agency was a vision of the promoters' goals; it was not what actually existed on Haulbowline when Sean Lemass 'turned a wheel' and the Krupp rolling mill went into production.

When the Minister spoke at the banquet in the Commodore Hotel, which followed the official opening, his speech was broadcast on Radio Éireann and fully reported in the press. It fell into two parts: that which he had prepared in the context of Irish Steel and the addition forced on him by the dire news which had broken during the day. For the industry itself and for the efforts of the directors he had unstinted praise. Their task had been enormous, fraught with obstacles and difficulties, 'lesser men would never have faced it or would have abandoned it undone'. . . it was his duty 'on behalf of the government and people of the state to assure Irish Steel of their goodwill'. However, Lemass had another message for his hearers, linked to the news that Hitler and his war chiefs were in all-night council and that Paris was to be evacuated. 'Only a miracle can avert war,' he said. 'A European conflict will, of course, stop our industrial expansion at once. . . . The widespread cessation of industrial activities with its consequent increase in the number of unemployed, the difficulty of obtaining supplies of essential commodities and many other problems which will arise once war begins, indicate that, apart altogether from the international or political difficulties which may arise, war will mean a period of great hardship even if our people never hear a shot fired. One virtue will carry us through it, the virtue of discipline, the willing acceptance of regulations which will inevitably be necessary for the conduct of affairs and voluntary public co-operation in making them work smoothly. I appeal to all those whose positions enable them to influence public actions in all walks of life to help to prepare the public mind for the coming situation.'

David Frame, in replying to a vote of thanks to him as chairman at the banquet, proposed by Seamus Fitzgerald and seconded by the Lord Mayor of Cork, is reported as saying, 'I hope I will not be alive when we start manufacturing munitions at Haulbowline. We want these great works for peace and I hope none of the present directors will live to see Haulbowline making munitions' (*Irish Press,* 25 August 1939).

These memories of a single day when Cobh was *en fête* for a new industry may seem overdrawn in a story covering so many years. They are, however, significant in underlining two elements, one being the Lemass commitment to the steel project and the other being the unfortunate conjunction in time of the opening of the new works with the circumstances of war. That the latter was the main factor in the subsequent failure is beyond doubt but it was not the only one. Even without a war, the promoters were aldready faced with an inescapable financial problem.

While the first steel bar made in Ireland was emerging from the last finishing stand in the rolling mill, Hitler's armies were poised to invade Poland and did so a week later. The declaration of war by Britain and France followed on 3 September. Ireland remained neutral throughout the war. This made it impossible to procure the materials and equipment which might have made the Haulbowline plant viable during these years. With difficulty some supplies of billets were obtained from Belgium while that country remained neutral but following the German invasion of the Low Countries in May 1940 both materials and equipment ordered by Irish Steel were seized and never reached Ireland.

In such circumstances David Frame's company was doomed to failure. It is remarkable that, in spite of all the difficulties, steel from the rolling mill was produced at all, however intermittently and at high cost, during the period from September 1939 until the appointment of a receiver in April 1941. The limited tonnage of steel produced at Haulbowline proved of considerable value to the national economy, particularly when used, as it was, to assist agricultural activities and as reinforcement for essential structures including defence installations. The need for independent steel supplies at a time when imports were restricted was one of the most potent arguments in favour of state assistance to the industry, indeed one which the Minister for Supplies, at least, weighed heavily against objections to that course.

It is unfortunate for the historian that the production records, costings and employment returns of Irish Steel for these critical years no longer exist. It would appear that, during the eighteen months prior to the first receivership, not more than 2,000 tons of finished steel were produced. From October 1942 when a new board of directors took possession from the receiver until the end of 1946, when the works again closed, a total of slightly over 8,000 tons was produced. During the whole period from 1939 to 1946 the level of employment varied with the scope of production, construction activities and maintenance. At its peak, the maximum number employed would have been 250 (including contractors' gangs) and then only for short periods.

In August 1939, as already remarked, the directors of Irish Steel were faced with a serious financial problem, no less formidable than the scarcity of raw materials. At that time the £450,000 raised by public subscription for completion of the steel works and for working capital had been almost totally expended. In September 1939, the first approach was made to the Department of Industry and Commerce for assistance in the form of a state guarantee of a loan of £250,000. This proposal was later formalised on the basis of revised estimates (prepared by Loewy) of capital requirements, mill outputs, costs and selling prices.

That form of assistance to industry was possible and had been used in many cases under the Trade Loans (Guarantee) Acts. The Minister for Industry and Commerce (with the consent of the Minister for Finance in each case) was the deciding authority and he was assisted in forming a view by reports from ad hoc Advisory Committees (composed of industrialists, economists and accountants) to whom applications were referred in the first instance.

When the Irish Steel proposal came to be heard, Sean Lemass was no longer in charge of Industry and Commerce. His portfolio as Minister for Supplies required all his attention because of the vital need to procure and allocate the essential supplies for maintaining the Irish economy and ensuring the protection of the community during the Emergency. While, naturally, he would have continued to be concerned about the possiblity of obtaining steel supplies from Haulbowline, the financial proposals for assisting the steel industry were a matter for his successor as Minister for Industry and Commerce, Sean McEntee, who had been in charge of Finance since 1932. His tenure of Industry and Commerce began on 16 September 1939 and ended on 18 August 1941, when Lemass resumed the role with which his name and reputation as a departmental minister are most often associated.

The application from Irish Steel was coolly received. It is evident that the advice given to the Minister was unfavourable to the company's proposal, mainly on account of the weakness of the security for the trade loan. There was also the factor that the additional capital was required for operations which relied solely on imported billets likely to be unobtainable as the war progressed.

A second application for assistance was made in February 1940. This time it was for a guarantee of a loan of £550,000 to include provision for the erection of open hearth furnaces, as in the original plan of 1938, so as to make use of native scrap. The estimates on which this second proposal was based were in the first instance prepared by Loewy, but a further

Kevin C. McCourt, Chairman of Irish Steel Ltd.

2. Former Irish Steel Chairmen

G.P.S. Hogan, 1956–74

J.C.B. MacCarthy, 1955–56

John Williams, 1947–54

3. View of Haulbowline *c.* 1870

4. Haulbowline in 1979

5. Visit of Minister for Industry and Commerce, Mr Sean Lemass, to Haulbowline, 1951

5a. David Frame

5b. Vincent Crowley

6 Visit of Taoiseach, Mr John A. Costello, with Minister for Industry and Commerce, Mr Dan Morrissey, 1948

7. Visiting group of National Association of Iron & Steel Merchants, 1952

8. Visit of Tanaiste and Minister for Industry and Commerce, Mr William Norton, 1954

and different set of estimates were later prepared for the directors of Irish Steel by H. L. Bengston of the International Construction Company, a London firm of consultants. Critics of the scheme fastened on the discrepancies between the various detailed assumptions and forecasts on which the company relied in seeking assistance. Moreover, the chances of obtaining materials and equipment for the furnaces still seemed remote.

It is evident that the Minister for Industry and Commerce and the Minister for Finance could find no solid ground on which public money could be risked in prolonging the life of the company and that they were prepared to allow events to take their course without any state intervention. The accounts and balance sheet of Irish Steel Ltd. for its first trading year disclosed a vulnerable situation with unpaid debenture interest and uncovered liabilities, a situation which caused increasing concern to the Munster and Leinster Bank. Although the industry at Haulbowline struggled on until the end of 1940, the inevitable closure took place in February 1941. The bank, as trustee for the debenture holders, appointed as receiver Vincent Crowley, the principal of the firm of auditors. He took possession of the works on 9 April 1941, when David Frame and his colleagues ceased to have control over the direction of the company. During this receivership, Crowley engaged in neither production nor constructional work, although he kept key supervisory and maintenance staff employed with bank credit. After an effort to sell the property, for which he received no bid, he concentrated on negotiations with the government directed towards finding some basis on which the industry could be revived.

The abortive attempts by the first directors to obtain state aid were revealed during a parliamentary debate about the Industry and Commerce estimates on 10 July 1941 (*Dáil Debates,* vol. 55, col. 1465 *et seq.*). Deputies Hurley and W. T. Cosgrave pressed the government to take some measures to reopen the industry at Haulbowline. In his reply Sean McEntee recounted the original proposals for a steel industry submitted to his department in 1937, the tariff protection granted and the basis of the capital issue floated by Irish Steel in July 1938. He described the first Trade Loan application for £250,000 (involving a change of policy by the promoters and the dependence on imported materials) which he had refused. The second proposal, to complete the original scheme would, as submitted, have involved new expenditure of £660,000 of which the state was asked to guarantee £550,000. The Minister said that in his view this was an underestimate even if it were possible to obtain the additional plant. He was critical of the competence of those who sponsored the company's proposals. 'The undertaking has closed', he said, 'because (*a*) the original plans as in the prospectus were not carried out and (*b*) the original management has not shown itself capable of running and developing this concern' (col. 1591). The Minister added that the manufacture of steel could not be resumed until there was certainty that plant could be delivered and that there was adequate technical staff to run it. Deputy Hurley maintained that the Minister's statements were exaggerated and indeed they seemed harsh and unsympathetic in the light of the promoters' difficulties caused by the war. One thing was clear, however, from McEntee's attitude. No state assistance was likely to be given without some change in the control and direction of the company. Worse than the Minister's strictures were the accusations by Deputy Corry, that the chairman of the company had delayed the construction of the furnaces because Hammond Lane could make more money during the war by exporting steel scrap to Britain than by selling it for melting at Haulbowline. (*Dáil Debates*, 7 July, vol. 88, col. 49/88).

The Minister for Finance (Seán T. Ó Ceallaigh) in December 1941 stressed the dangers of involvement and asked the government to obtain an independent expert view on the economic possibility of an Irish steel industry before commitments were made. Again in March 1942 he urged upon the government that the only safe course was to avoid any Exchequer commitments in respect of Irish Steel Ltd. The officials in Finance were tendering advice which reflects the apprehensive view traditional in treasury corridors. The author recalls, although memory is fallible, the conclusion, 'there is nothing in this project but the certainty of loss and the possibility of discredit' written by one prophet!

Reconstruction of Irish Steel

After much parley between the receiver, the bank and

the government departments concerned, a project for assisting the industry to reopen was put together within the provisions of the Trade Loans legislation. The scheme, in the government's view, was a non-starter unless the new capital was to be applied in the main to the erection of steel-making furnaces, and unless the bank agreed to share its existing security with the state. The Minister for Industry and Commerce was, as a condition of his guarantee, to be given power to appoint all the directors of the company for so long as the Trade Loan remained unpaid. Creditors of the company were to be offered debentures in lieu of cash, such debentures ranking after those held by the Minister and the bank. The existing directors had previously offered to abdicate and did so at a general meeting of Irish Steel Ltd., which made the necessary changes in the Articles of Association of the company, thus leaving the Minister or his nominees in control of the industry and effectively depriving the shareholders of their rights for as long as the state guarantee was outstanding. One difficulty remained, the possibility that existing creditors who had enforceable contractual rights might drag the company into liquidation and bring the scheme to nought. Some of these creditors, notably Loewy, Huxley, the managing director and McCarron, the secretary, who had service agreements with the company, refused to accept the third debentures. As a precaution, at the instance of Sean Lemass, the government took the unusual step of depriving such persons of their legal rights by an Emergency Powers Order (E.P. no. 194, Order 1941), not, of course, as a permanent bar but as a temporary arrangement while the emergency conditions lasted. This action by the government was hotly challenged as unjust abuse of power by opposition speakers in both the Dáil and the Senate, but motions to annul the Order failed. Lemass put the matter plainly, 'This is in my view an essential step in the measures I am taking. Would you prefer the steel works to remain closed'?

In the Senate, Sir John Keane asked why did the government not come in, buy out the company and start a second company? The Minister pointed out, 'in such circumstances the unsecured creditors would get nothing at all. What I am doing with the Trade Loan is giving them a chance of getting something in the future' (*Seanad Debates* vol. 26, col. 2253).

In October 1942 the guarantee of a Trade Loan of £200,000 (to be advanced by the Munster and Leinster Bank) was approved. The loan together with one half (£125,000) of the original debenture stock was secured by a first charge on the company's assets. The other half of the stock was secured by a second charge and the remaining creditors had to be content with third ranking debentures of which about £42,000 was issued.

It was part of the arrangement that the arrears of interest due on the original £250,000 of 5½% debenture stock (about two and a half years' interest of £35,000) would be paid from the Trade Loan, i.e. to the Munster and Leinster Bank as trustees for the debenture holders. The only other pre-reconstruction liabilities to be met from the loan were the receiver's expenses. The rest of the £200,000 was expressly earmarked for the construction of two open-hearth furnaces and for working capital to enable the company to recommence production and trading.

The new directors of Irish Steel, appointed by the Minister in September 1942, were Vincent Crowley (chairman), M. Osterberg, A. O'Rahilly, C. Murphy and J. Gannon. Crowley, the former auditor of the company and later receiver, was a director of Irish Tanners Ltd. and in his professional capacity had many important assignments including the national electricity and air transport undertakings. Osterberg, a Dane, was managing director of Cement Ltd. Murphy was a leading Cork textile manufacturer. O'Rahilly, a prominent industrialist and engineering graduate, had successfully pioneered his own plaster-board industry, Weatherwell Ltd., and was a director of the Turf Development Board, a state enterprise. Gannon was also an engineer and a director of AET Ltd., the electrical manufacturers in Co. Louth. On their past records and achievements, these men formed a strong board, although naturally, with the exception of the chairman, they had no previous experience in a steel industry. One of the directors, Con Murphy, resigned either late in 1943 or early in 1944 and was not replaced. The other four directors marched on bravely until the end of the road in January 1947.

In the national sense, Aodhágán O'Rahilly was a leading figure in the reconstructed board. Son of The O'Rahilly, who was killed during the fighting in Dublin at Easter 1916, he too had fought for his country. Intensely devoted to the cause of Irish industrialisation and of economic independence, he would have spared no effort to carry out successfully the difficult assignment at Haulbowline so far as a part-time director could influence events.

The re-organisation did not provide for a managing director. This placed a greater burden on the part-time board, all busy men in other spheres. The board met monthly at Haulbowline and also held occasional meetings in Dublin for urgent business. Crowley gave considerable time to the company and, according to O'Rahilly, was virtually acting as executive chairman. Alfred Donohoe, an able accountant from Crowley's firm, was appointed secretary to the company in place of McCarron who had resigned, and was the chief administrative officer. The technical team was led by Maurice Lambot, a Belgian, who appears to have held a dual position as company executive and, through his firm Lambot Frères, as designer and installation contractor for the two open-hearth furnaces. In February 1943 production from the merchant bar mill recommenced. Lemass explained the position of the industry to Dáil Éireann on 20 May 1943 (*Dáil Debates*, vol. 90 col. 472). Production was being carried on with steel billets and rails accumulated by the receiver during the closure and, in view of the limited output, it had been decided to confine sales from Haulbowline to essential purposes such as horse shoes, bolts and nuts, nails and agricultural machinery and implements. The Minister was pessimistic about the company's ability to erect the furnaces because necessary equipment could not be imported.

Due to Vincent Crowley's position of chairman, it was necessary to appoint new auditors to Irish Steel. Purtill and Co., a Dublin firm, were chosen. In December 1943 the auditors submitted a report and accounts for the three financial years ended 31 March 1943 and a balance sheet at that date. The first directors of the company had not filed any accounts or auditors' report subsequent to those relating to the period ended March 1940.

The statutory returns made by Irish Steel Ltd. to the Companies Office (Public Record Office, file 9606) give some indications of the company's financial position from

1942 onwards. At 31 March 1943 there was an accumulated loss of £93,000 on the profit and loss account. The company's financial year 1943–44 showed a loss of £82,000; for 1944–45 the loss was shown as £76,000; and for the last year of Purtill's audit, 1945–46, the loss was £132,000 with outstanding capital liabilities under first mortgage debentures of £481,172. The certified balance sheet of liabilities and assets at 31 March 1946 at a total of £765,000 approximately, showed clearly that the company was insolvent. No company records have survived to show, for any part of its trading history, the volume of outputs, turnover or selling prices. (A description of the production line at the end of 1946 is included in a note at the end of this chapter.)

There is however, a departmental note stating that 8,000 tons of steel were manufactured during the period from March 1942 to August 1946. From the same source there is information that 3,500 tons of Canadian billets had been purchased by Irish Steel for £73,000 (exchange control approval for the dollar transaction having been given during the last stages of the war) but it is not clear whether all of this tonnage reached Haulbowline. It was noted that the bars rolled from these billets had to be sold by Irish Steel at a substantial loss because of the maximum price order imposed by Sean Lemass which applied both to imported and home-produced steel. After the war British steel was, for a time, offered in Ireland at £20 a ton ex-duty and this competition as well as the controlled selling price increased the Irish company's difficulties. At periods during the Emergency, Irish Steel had been able to sell at a profit. Indeed, when there was no price control Aodhagán O'Rahilly recalls 'we had sold rollings at £100 per ton'.

In July 1946 according to a note then written by Tom O'Shea in Industry and Commerce, the company's wages bill was £500 per week. This suggests a lower level of employment than has been assumed from other evidence.

The talented directors of Irish Steel whom Lemass had appointed in September 1942 cannot be blamed for the second failure of the company. Nor indeed can the staff who struggled vainly against lack of materials and inability to secure delivery of the furnace components until too late in the day to avert insolvency. At one stage during the war, according to George Radley, Lambot made a gallant but futile effort to construct from available materials a small cupola furnace, in which to reduce cast iron scrap to molten iron and a converter to refine the molten iron into steel. It was not until May 1946 that the open-hearth furnaces were completed according to Lambot's design, but reasonable production results were not secured until John Davies, from the British steel industry, took charge of the steel-making operations while, at the same time, Percy Austin was recruited to supervise the rolling mills. Their arrival appears to have coincided with the departure of the Belgian technical command who had been there since David Frame engaged them for furnace and gas producer installation. Lambot and his assistant Moucheron, who had been in charge of quite a large design and drawing office at Haulbowline, left behind them a number of Belgian tradesmen whose employment at the works continued until 1948.

It is impossible in the absence of company or works records to reconstruct the events of the frustrating years 1942–46. Aodhagán O'Rahilly has recalled some of the incidents in a letter to the author with permission to quote him:

> Our job was to get the merchant mill into production and roll rounds, flats and angles from any available raw material (mostly the bulbs of old rails, I suppose cut off with a burner); and also to get an open-hearth furnace into operation, so as to use native scrap . . . we got the mill working — it was a Krupp mill which Frame had picked up for maybe £50,000. It was a first class piece of machinery and indeed was the only machinery in the place which was not scrap or near it.

As regards raw material O'Rahilly refers (as Lemass had done in a statement to the Dáil in October 1942) to some billets in a ship torpedoed off the Donegal coast: 'On a calm day at low water it was possible to stand on the cargo of billets and still have your head out of water. Each spring as calm weather and easterly winds were anticipated, a local merchant would come to us and bargain with us as to what price we would pay for the several thousand tons of three and four inch billets, twenty feet long, which he was hoping to salvage.' It sounds like a romance similar to *Whisky Galore*.

Some tonnage was indeed received by Irish Steel from this source but it was not a lot in O'Rahilly's recollection.

O'Rahilly recalls asking Maurice Lambot, whose firm (transferred to Birmingham after the invasion of Belgium) had the furnace installation contract, whether he would build the furnaces at a fixed contract price, and receiving the reply, 'Mr O'R. I may be a fool, but I am not a bloody fool.'

The great trouble was to get export licences from the British authorities for the essential material, steel work and bricks necessary to complete the furnaces and also for coal to feed the producer gas plant. The British did, during the war, give small allocations to Ireland of coal and steel required to maintain food production. O'Rahilly records 'the coal was allocated by the Coal Controller and he said he had already given Ireland its ration and he was not concerned with steel. They were arguing about it six months later when we got the furnace materials from the US or Canada. It was a classic example of too many cooks spoiling the broth.'

It is evident that O'Rahilly played a prominent role in the efforts to secure material for the furnaces. In the winter of 1944 he and Vincent Crowley visited London with their buying agent (probably Darby and Company, who had had an agency agreement with Irish Steel since David Frame's time) in order to put pressure on the British authorities. It was argued that if the Irish Steel furnaces could be built, not only would the British be spared any allocation of steel for Ireland, but a surplus could be exported to Britain. O'Rahilly recalls wartime London at night when air raids were a constant hazard; the duties of the directors in those days seem certainly to have been more dangerous, if not more difficult, than those imposed on their successors in the industry. Strangely enough O'Rahilly was not insured by the company against accident during this trip, although apparently his chairman had taken that precaution. 'Presumably I was expendable,' was his facetious comment on the omission.

Aodhagán O'Rahilly ends his recollections by a reference to what he terms 'the swan song': 'Lemass was going to make an order of some kind which would have saved Irish Steel. I suppose I heard this and went to see him. He was working on the order when I went in to him and I urged my opinion which was to scrap the rescue attempt and to go into voluntary liquidation. When Lemass heard my views he said "I think you are right" and picked up the paper he was writing on and threw it into the waste-paper basket. Then I had to go back to my colleagues and tell them I had given the *coup de grace* to Irish Steel. I may add that I subsequently regretted what I had done and believed I made a mistake.'

O'Rahilly's account of his interview with the Minister, probably late in 1946, invokes a mystery. There is no other evidence that an 'order' of any kind – the only kind possible was one under the Emergency Powers Act – was being contemplated at that time. And if it was, for what purpose? Possibly the Minister was considering a further rejection of the bank's request that a receiver be appointed and he may have been writing a direction on that point when O'Rahilly settled his mind in favour of prompt liquidation and a new start under state auspices. Sean Lemass, a realist, must have known at that time that a receivership could only be postponed. There was no hope for the industry's survival if it had to carry its existing load of capital debt on such small productive capacity as the then installations allowed.

In any event it is on record that the Minister for Industry and Commerce met the four directors of Irish Steel on 17 December 1946. Before that date Crowley had reported to the department that he had run out of cash, was under severe pressure from the bank and could not keep the company in business beyond January 1947. It is probable (but this is surmise) that on meeting the four directors, Lemass told them of the failure of efforts made through the Industrial Credit Company to achieve, with the bank's consent, a capital reconstruction of the company which would involve a drastic writing down of the rights of shareholders and debenture holders. He had decided to join with the bank in placing the company in receivership forthwith, in the hope that the assets could be sold as a going concern with continued operation of the industry by the purchasers. Doubtless the Minister also thanked the directors for their sustained efforts to maintain the industry in face of insuperable difficulties.

Vincent Crowley was once again appointed Receiver. He took possession of the company and its assets on 8 January 1947. The outcome of the receivership is related in the following chapter.

The file in the Public Record Office (9606) contains a letter from Vincent Crowley dated 26 September 1949, informing the Registrar of Companies that he had ceased to act as Receiver or Manager of Irish Steel Ltd. Subsequently the Registrar addressed numerous enquiries to 'the Secretary of Irish Steel Ltd.' as to whether the company was carrying on business. There was no reply. One may guess that the officials of Irish Steel Holdings Ltd. were unwilling to accept the slightest identity with the former company at Haulbowline, a necessary posture in face of claims from shareholders and creditors.

Lemass made the position plain in the Dáil. 'ISH has no connection with Irish Steel Ltd. and has no responsibility for any liabilities of the latter company' (*Dáil Debates* 21 February 1952, col. 902). Eventually the patience of the Registrar of Companies was exhausted. He gave notice in the *Official Gazette* of 3 October 1952 that Irish Steel Ltd. was dissolved and had been struck off the Register of Companies. This formality legally ended the existence of the company which David Frame and his associates had launched fourteen years earlier with public fanfare and high hopes and which, after their failure, had been controlled by other directors for most of the company's effective life. As events turned out the demise of Irish Steel Ltd. was not allowed to bring the steel industry to an end.

Note on production at Irish Steel Ltd.

Mr George Radley very kindly furnished the author with an account of the operations at Haulbowline during the last period of Irish Steel Ltd., prior to its closure in November 1946. He was, at the time, employed in the engineering department of the company. His recollections and comments have historical value and hence they are reproduced verbatim:

Prior to the close the following was the situation. Two Lambot-designed open-hearth furnaces, each of 35 tons capacity were in operation. Raw (i.e. uncleaned) producer gas was the firing medium and was generated by 2 large gas producers of Lambot design. Both of these furnaces tapped into the 35-ton Ladle Car which ran on rails in the middle of the Casting Bay Building. This car then teemed the steel into clusters of small ingot moulds of 3¼ cwt capacity. Each cluster had a total of 24 moulds radiating in 6 branches from a central trumpet feeder. On average 220 ingots were teemed from each 35-ton cast.

These ingots were stripped and sorted in the Ingot Bay Building. This building houses the soaking pits at the present time.

The new 35-ton open-hearth furnaces should have been capable of producing 35,000 tons in a year but such a figure was never reached.

The Lambot furnaces and gas producers could be considered standard production units for the period, and the blame for erratic performance lay mostly with the handling equipment. Wharfside cranes, furnace chargers in particular, and all the overhead cranes were, indeed, in very poor mechanical and electrical condition and breakdowns were occurring too frequently. There was no stand-by crane equipment whatsoever and when one unit went down, the whole production also had to close down.

The adoption of a small ingot of 3¼ cwt was intended to avoid the necessity of using a blooming or cogging mill. The 6" x 6" ingot section could be reheated in the Krupp 'billet' reheating furnace and rolled directly into rounds, squares, flats and angles in the Krupp merchant bar mill. The smaller sizes had, however, to be double rolled i.e. billets would be produced from the ingot and later these billets would be reheated and rolled into half-inch and three-eighth inch bars.

The small ingot scheme had the advantage that it avoided the cogging mill expenses. It had increased feeder (refractory) brick costs in the ingot making but these were not excessive. The Belgian Clabecq steelworks were producing steel by this method up to ten years ago and had the reputation at the time of producing the cheapest reinforcing bars in Europe.

The transporting of ingots to the merchant bar mill was carried out by means of a horse-drawn rail car running on Jubilee rails from the Ingot Bay to the Merchant Mill Annexe building. At the time of the 1946 closure this system was still in operation. [ISH board minutes record the sale of horse, harness and cart for £25, tractor/trailer transport having been substituted. *Author's Note.*]

The Krupp merchant bar mill consisted of a 10-ton per hour coalfired reheating furnace, a three stand 3-high 430mm roughing mill, a five stand 3- high 350mm finishing mill and a mechanical cooling bed and cutting up line. The mill also had four 2- high 250mm wire mill stands and two rod coilers for the production of wire rod in coiled form.

The billet furnace had to be altered in skid construction in order to accommodate the small ingots of 6" x 6" x 48". Two row pusher charging was adopted and a manually operated overhead monorail tong suspension unit was used to discharge the heated ingots from the furnace.

The 3¼ cwt ingot had to be counterbalanced on the suspended tongs by the weight of a man. This man actually straddled on top of the tongs at one end, the ingot at the other end and a second man pushed both around the furnace. In this manner heated ingots were delivered to the approach roller track of the roughing stands.

The operations at stands one, two and three of the roughing mill were somewhat similar to the present mill arrangement with tilting tables and roller tracks undertaking the heavier operations for the tongsmen.

The five 350mm stands of the finishing mill were in line across and four mill operatives were engaged, two front and two back, hand looping the bar through these mill stands.

Repeaters at the back were used occasionally but these were not very successful and good hand looping by men was deemed the better way to achieve output.

As mentioned earlier, billets had to be produced from the small ingots in order to roll three-eighth inch and half-inch bars. These billets were rolled on the 430mm roughing stands and were cooled on a simple skid rail cooling bed and afterwards moved manually across into the Annex Building for storage. The actual rolling of three-eighth-inch rounds from these billets was unusual in so far as a double loop was required at stands one and two of the finishing mill. The bar, on entering stand one, was hand looped at the back into stand two and on emerging at the front was hand looped back again into stand one, after which it looped again into stand two and then on to three, four and five stands.

This double loop operation was so dangerous that an automatic double repeater had to be introduced at the back of both stands. The front loops had, of course, still to be manipulated by hand and the loopers had narrow escapes from time to time carrying out this particular operation.

The cooling bed and cutting line of the mill were reasonably good installations and only the Krupp rotary shear to the bed could be faulted for holding up production. The product range of the mill was as follows:

Rounds: ⅜" to 1½"
Squares: ⅜" to 1½"
Flats: ¾" to 3"
Angles: 1" to 3" equal angles.

Prior to World War II, the mill produced bucket and churn iron and also heel tips for army boots. During the Emergency rounds and angles were produced from double bulb and flanged rails; railway axles; and railway wheel rims.

Merchant Bar production using normal billet raw material should have yielded 16,000 tons per annum on single shift and 32,000 tons per annum on double shift. The small ingot techniques would reduce these figures by a third. The mill, however, never achieved consistent production with the small ingot technique and the three-eighth-inch and half-inch bar sizes were always in trouble. The three-eighth-inch was abandoned before closure.

G. Radley, December, 1975

CHAPTER V

The New Departure

Sean Lemass broke a long parliamentary silence about the affairs of Irish Steel Ltd. when, on 22 April 1947, he presented to Dáil Éireann the estimates for the Department of Industry and Commerce for the financial year 1947–48. Although there was no provision for the steel industry in these estimates, the Minister thought it necessary to explain a change of policy and the circumstances that caused it. The board of the steel company, he said, whom he had appointed in 1942, had on many occasions asked to be relieved of their responsibilities and the company had, with his agreement, been placed in the hands of a receiver. The latter had obtained no offers for purchase of the industry as a going concern. The only other offers had been in respect of portions of the plant with the intention of exporting them. If this were done, Lemass said, it would involve selling the residue as scrap and the termination of the industry. 'This I do not wish to happen and it has now been decided to form a state-financed company to acquire all the assets from the receiver' *(Dáil Debates* vol. 105, col. 815). There were few comments on the Minister's announcement. Deputy Coogan said that if it meant entrusting the venture to civil servants, whose training and functions unfitted them for commercial tasks, he was against such an expedient. Naturally Deputy Corry, while not committing himself to any observations regarding civil servants, wholeheartedly favoured continuance of the industry in any fashion. During his long tenure as deputy for East Cork, Martin Corry never failed to act as vexatious prodder of ministers and directors alike in all matters concerning production and employment in the steel industry which so closely affected his Cobh constituents.

The Minister's public statement on 22 April was the sequel to discussions which had been going on since the beginning of the year both between departments and at government level. These culminated in a decision towards the end of February that, if the receiver failed to sell the undertaking as a going concern, a temporary company should be formed by the Minister for Industry and Commerce to acquire the assets. The emphasis was on the temporary nature of the expedient as a means of reviving production at Haulbowline during the period of post-war steel shortages. There was no intention that the decision could imply any permanent state involvement in the steel industry. Many thought it would be more expedient to subsidise the existing company since a subsidy could be more conveniently terminated than a fully-fledged state enterprise.

Although unwelcome it was no surprise when Vincent Crowley informed the Department of Industry and Commerce on 21 March 1947 that he had received no bids for purchase as a going concern. He added that, as Receiver for the debenture holders, his duty would be to obtain for the assets whatever price he could get from a break-up sale. The chances of selling the industry as a going concern had always been remote. At that time there was no Industrial Development Authority nor were there the concomitant grants and taxation reliefs which might have induced foreigners to invest in an Irish steel enterprise. Even had these advantages existed, it was unlikely that the pressures of post-war reconstruction in Britain and other European countries would have permitted anyone to consider seriously a risky acquisition in Ireland, however cheaply the property could, in the circumstances, have been bought. As for the private sector at home, the sad record of the existing company and the loss faced by its creditors and stock holders were ample reasons why no Irish firm or consortium could be found to provide capital for a renewal of the effort to establish a viable steel industry at Haulbowline.

Thus the dilemma facing the government was either to allow the steel industry to disappear or to put into immediate effect the decision already provisionally taken at the end of February. Sean Lemass was essentially a pragmatist who did not allow his judgment to be unduly affected by precedents or by hitherto accepted views regarding the proper domain of state action. While the collapse of the steel industry was a relatively minor matter in the context of the many social and economic problems of 1947, it was relevant to one anxiety, that of rising unemployment and emigration. There was also the point, to which the Minister gave importance, that national security called for the production of steel. It was, in his view, manifestly desirable to avoid in some new Emergency, not unlikely at that time if the East-West confrontation became bellicose, a repetition of the failure to produce even a modest proportion of the steel required for essential activities in Ireland. These considerations had caused Lemass to press upon the government the need to purchase the assets from the receiver and to operate the industry as a state agency. One may well suppose that his official advisers counselled caution. The investment was a high risk, and that commercial failure was not unlikely could scarcely be regarded as an over-pessimistic view.

The Minister for Finance, Frank Aiken, having recently paid £192,000 in respect of the guarantee of the Trade Loan of 1942, was ill disposed towards a risk of further Exchequer loss in attempting to save the industry. Moreover, during the decade following the war, considerable capital expenditure would be needed for the building of schools, hospitals, housing, airports, harbours and many other constructional works requiring steel. To add to the cost of these operations by maintaining under tariffs and subsidies a native steel industry producing (as it was contended) a product much dearer than imported steel would increase Exchequer burdens and also penalise the private sector.

It was after reflection on all these diverse considerations that the government took its decision in favour of acquisition and, faced with the Receiver's warnings of 21 March, Lemass moved rapidly to implement it. Vincent Crowley was told to defer action as a bid for the assets would be made. In agreement with Finance, Lemass decided that a 'working party' of three civil servants should negotiate with the Receiver, having authority to purchase the assets for a sum not exceeding £250,000, and that, if successful, they should form themselves into a company to complete the transaction. As a company they should operate the industry commercially and as soon as they had sufficient experience of the operations and the problems of the industry, they should furnish a report with recommendations to the Minister for Industry and Commerce. The question of working capital for the new company was left for solution by the working party after they had acquired ownership.

The three civil servants appointed to assume this novel responsibility were all busy officials, Williams and O'Shea in Industry and Commerce, and Hogan in Finance. It was a part-time assignment to be taken in their stride as complementary to their existing jobs. As a working party with the mandate already described, they met on 2 April 1947 to consider their plan of campaign. Because the outcome depended to some extent on their backgrounds, on their approach to the task and on their collective determination to succeed, some account seems necessary of the three individuals who, so fortuitously, became part of this history.

John Williams, then in his middle fifties, had been one of an exceptionally talented group of boys who had gained exhibitions and scholarships at the North Monastery school in Cork. By open competition he secured a post in the second division (executive) of the British civil service and was assigned to the Irish establishment. That was in 1913. He rose steadily, if unspectacularly, in Dublin departments until he came into prominence during the Emergency as controller of many vital commodities in the Department of Supplies. He returned to Industry and Commerce in August 1945 as deputy secretary or second-in-command, dealing with industrial development while still retaining control of the rationing and pricing systems of the early post-war years. A man of forceful character and somewhat brusque personality, capable of rapid decisions, he was well-suited to the sort of responsibility the steel enterprise demanded. Moreover he was no novice in dealing at the highest level with business people in the private sector.

Tom O'Shea, from Kilkenny, and six years older than Williams, began in 1908 as a junior clerk in the former British civil service in Ireland and following wide experience in several departments, he found his senior administrative niche in the war-time tasks of the Department of Industry and Commerce. In 1947 as assistant secretary under Williams, he was dealing officially with many industrial problems including those of the steel industry. Perceptive and fair-minded, with a vast store of practical common sense, O'Shea also had much personal charm. His gift of getting on with people and of mollifying opposition proved to be a considerable asset when dealing with those who disliked or distrusted the idea of placing control of a business venture in the hands of 'a bunch of civil servants'.

The third and youngest member of this triumvirate, G.P.S. Hogan, came into the affair as the financial watch-dog. A university graduate and a barrister by profession, he had been recruited in the early days of the Free State under the 'public interest' clause in the civil service code. In 1947 he was assistant secretary in the Department of Finance in charge of the Irish Exchange Control, as he had been during the whole period of the war. Like Williams he had taken part in the inter-governmental trade and payments negotiations during the era of bilateralism which immediately followed upon the conclusion of the war. His exchange control duties had brought him into direct personal contact with many aspects of industrial and commercial life, both national and international. Hogan had been, since 1939, a non-executive director of the state-owned industrial alcohol distilleries and, indeed, had recently been joined on that board by Williams as the nominee of his Minister.

It was not unusual for senior civil servants to be appointed to the boards of state-sponsored undertakings. Pre-eminent among such was John Leydon who, while retaining his full duties as permanent secretary of Industry and Commerce, was also the chairman of two state transport companies, aviation and shipping. In such cases the boards were mainly composed of men from the private sector of business. The rescue operation for the steel industry provided the unusual feature that civil servants alone comprised the directorate. This was inevitable in the circumstances of haste and risk and, in any event, it was designed as a temporary arrangement.

The working party lost no time. Their first task was to secure an independent valuation of the assets and a report on the condition of the plant at Haulbowline. Through the good offices of the British Board of Trade they were able to obtain the services of E. D. Bool, a recently retired steel-maker from Sheffield. They also asked Col. J. J. Winters B.E., then the official arbitrator, to value the leasehold interests and the buildings. The working party themselves, on the basis of the information on the departmental files and on the returns made to the Prices Division by Irish Steel Ltd., made calculations as to the ceiling of a purchase price which could be paid with any hope of making steel at economic prices. When Bool and Winters had inspected the property and furnished their reports it was evident that earlier views of the market value of the assets were erroneous. A contemporary note by Hogan states 'these reports had an important bearing on the subsequent negotiations because they clearly show that not only was £250,000 (the limit fixed by the government) much too high as a capital investment for the purchase of the fixed assets, having regard to the necessity to provide also for repairs and renewals and for working capital during a period of trading losses, but also that the break-up value of the assets was much less than might have been assumed from the balance sheet values of these assets.'

On 16 May 1947, the working party met to consider their advisers' reports and the offer they might make for the fixed assets. They had with them Dermot McDowell of John M. McDowell and Company whom they had appointed their solicitor to effect purchase and to take steps to form a company. It was decided to convey to the Receiver a written offer of £75,000. This he immediately rejected as entirely inadequate, at the same time requesting permission to dismiss the key staff and to proceed at once to a break-up sale. On 24 May, in a long discussion with Vincent Crowley, the working party advanced their offer to £100,000.

The working party were in a position to drive a hard bargain and they did so. They assumed that Vincent Crowley could not have got from his advisers, the International Construction Company of London, any valuations for break-up sale which were seriously at variance with the Bool estimates. Accordingly, when Donovan, general manager of the Munster and Leinster Bank, attended the final meeting with the Receiver on 3 June 1947 as trustee for the first debenture holders other than the Minister, agreement was reached on a figure of £125,000. This price was to include the liquid assets on site (stocks of finished and semi-finished steel, scrap, coal, etc.) which, under the Terms of Sale, the purchaser of the fixed assets was bound to take over at cost valuations certified by the Receiver.

Odd though it may seem, the Receiver required the formal approval of the Minister for Industry and Commerce before he completed the sale, even though the purchase offer had been made by the Minister's own agents. This was so because the Minister jointly with the Munster and Leinster Bank, as owners of the first debentures, had appointed the Receiver. In his letter to the Minister requesting approval, Crowley stated that the bank had already approved and 'I expected to realise much more than this sum (£125,000) for the assets but it is my considered view that no higher sum can be obtained and that the offer should be accepted.' To this recommendation Sean Lemass agreed without delay although, as a keen card player, he may momentarily have wondered what might have occurred if the three civil servants, holding the hand they had, had found themselves called upon to make one more bid.

Williams had conducted the negotiations with firmness and skill, judging accurately the lowest figure which the Receiver and the bank would be willing to accept for a quick sale. It was also within the figure which the working party had concluded would be the highest price that ought to be paid if the new venture was to have a chance of commercial success. The working party thus found themselves, in trust for the state, the legal owners of a property they had never seen and in control of an industry about which their knowledge was confined to information on departmental files concerning the melancholy experiences of the former company. Money to complete the purchase had to be obtained speedily. Since the government had no parliamentary authority to provide Exchequer funds for the purpose, bank finance was necessary. The Bank of Ireland provided facilities to enable the deposit required by the Receiver to be paid immediately. This accommodation was soon increased when the Minister for Finance guaranteed borrowings by a company entitled Irish Steel Holdings Limited.

That name had been given for want of a better one by the working party, with the approval of Sean Lemass, to the company incorporated on 23 June 1947. The term 'holdings' was chosen to indicate what was hoped to be a temporary stage pending permanent arrangements, possibly in association with commercial interests if the new venture should prosper. The company was a private one with an authorised share capital of £100 of which three shares were issued. Williams, O'Shea and Hogan each paid over his pound note for the privilege of subscribing to the Memorandum and Articles of Association and being the first directors of the company.

The three civil servants took considerable care in drafting the Articles for the approval of the Ministers. Shares could only be transferred with ministerial consent and no dividends could be paid. All powers of appointment of directors, of approving their remuneration, of consenting to borrowings, and so forth, were vested in the Minister for Industry and Commerce, acting in financial matters, with the consent of the Minster for Finance. But the day to day control of the company belonged to the three directors, a responsibility which, in general terms, was to be exercised without consultation with Ministers or their departments. Lemass was emphatic that this procedure should apply comprehensively. In an interview when Williams and his colleagues reported on the completion of the purchase, the formation of the company and the taking over of the steel-works and the personnel at Haulbowline, the Minister wished them good fortune in bringing the plant into production and added 'I don't need to hear from you again unless you are in trouble'. This was a characteristic Lemass exercise in delegation and in keeping with his belief that a state company should operate commercially and that its directors should have effective responsibility and power.

As Williams was ill, it fell to O'Shea and Hogan to take over the works from the Receiver on 4 July 1947. They inspected the idle plant which provided a depressing aspect, matched by the dejected demeanour of the key personnel. These were about a hundred in all, administrative, supervisory and labouring, whom Vincent Crowley had continued in employment during the receivership so as to maintain the equipment and present the appearance of a going concern. The directors addressed these people, explaining the new departure and preaching hope for the future. The response was unenthusiastic, indeed incredulous, so it was felt necessary to show teeth, much after the fashion of the shooting of Admiral Byng on his quarter-deck *pour encourager les autres.* The individual selected for sacrifice was the sales manager whose function in a business that had not produced for many months was obscure. He was given a month's pay in lieu of notice. It was embarrassing for O'Shea and Hogan to meet the same gentleman later in the day at a place of refreshment in Cobh but he silenced the stammered regrets by saying, 'To tell you the truth, I'd have thought you were mad if you hadn't done it.' Bob Cogan set up his own business as a manufacturers' agent, in which capacity he rendered much service to the new company and indeed became a good and valued friend of the directors, who had so summarily dismissed him at their first meeting.

Before and during the negotiations with the Receiver, the directors had discussed amongst themselves how they ought to proceed in the event of their bid being successful. They were all busy officials and could not give continuous detailed attention to the industry. Their attention would, in any event, prove ineffectual unless it was, at the outset, given on the spot at Haulbowline. A deputy was needed. They decided to rely on the resources at their own command. There was on O'Shea's staff a junior officer whose relative inexperience was offset by other qualities which made him a front runner for this particular task. Christopher Fitzpatrick, although born in Cobh, grew up on Haulbowline Island where the family home was during the years of involvement in the dockyard. Fitzpatrick's local knowledge and acquaintance with the Cobh community were distinct assets. So also were his technical abilities, since he had qualified as a mechanical engineer at the Crawford Institute in Cork. (Some years afterwards he proved his versatility by graduating with honours from the Commerce Faculty in UCD).

After some service as a vocational teacher in engineering subjects, Fitzpatrick obtained, in 1937, a post as factory inspector in the Department of Industry and Commerce. At his own request he was later transferred to the administrative side of the department and thus in 1947 he was a higher executive officer in the Trade and Industries Division but not concerned directly with the steel industry and its problems.

The directors obtained Fitzpatrick's services on secondment from the department to the new company and, overcoming his reluctance, they transferred him from Dublin to Haulbowline. His initial mandate was to report on the condition of the plant and to take such steps as were necessary to resume steel production as soon as practicable, in accordance with the policy and instructions of the board. Fitzpatrick's first appointment, for these purposes, was described as 'Assistant to the Directors' but in a short while, as matters progressed satisfactorily, he was given the status of general manager, a position he retained until his departure from the company in December 1955. As a seconded officer, he was paid by the department according to his rank as HEO. The company reimbursed his salary to the department and paid him additional allowances and emoluments, as determined by the directors, in respect of his services as general manager.

For the job abruptly thrust upon him, Christy Fitzpatrick had an array of formidable attributes, the same which in later years made him so successful in his own business. Possessed of ambition, energy, a buoyant temperament and courage, when facing difficulties and delays, fertile in expedients and stratagems, confident in manner and speech, he rapidly breathed new life into the moribund industry. His engineering background and skills gave him an advantage. He was at home on the factory floor and could speak effectively to the men. It is said, too, that he could handle a tongs or drive a crane, or deal with an electrical fault if the need arose. It was a considerable achievement that, in the space of three months, such progress was made in reconditioning the plant and establishing work procedures that steel was produced and sold before the end of September 1947. This was no false start. The operations were continuous and output and costs were far better than any achieved by the former company. During this critical period Fitzpatrick received much encouragement and support from his chairman, who kept in constant touch with the works.

It was no surprise that Fitzpatrick eventually found his true métier in private enterprise. Apart from his own ambitions, his somewhat autocratic disposition must have found irksome the extent to which his directors kept a close rein on all company activities during the formative years. Recalling these events nearly a quarter of a century afterwards, Fitzpatrick said that he could never have remained happy running a state industry, where in his view a chief executive's position was insecure and vulnerable to ministerial pressures. When he resigned to embark on an independent career he was well equipped to do so, not only by his own talents but also by the experience of industry and management that a lucky chance had given him the opportunity to acquire during his exacting years of public service at Haulbowline. His engineering and trading business in Dublin, now a group of many companies became, in due course, a major customer of the steel company he had served in its early days 'with ability, initiative and singleminded devotion', to use the words of Williams at the first AGM on 9 October 1948.

Only a few months after production at Haulbowline had begun and while many problems faced the directors and their general manager, other circumstances arose which might have called a halt to what has been termed 'the new departure'. The revival of the steel industry in public ownership had been a political decision taken by Sean Lemass, to which he secured government assent. The measures taken to implement that policy were executive actions directed by him and they had not been the subject of any parliamentary approval in the form of vote or resolution, apart altogether from legislation. What had been done, therefore, was liable to revocation, or at least suspension of activity, if a change of government should take place. Such a change did happen in February 1948 when a general election ended sixteen years of Fianna Fáil government and of the powerful influence of Sean Lemass in economic affairs. The new administration was described as an Inter-Party government (the term 'coalition' was avoided) and was composed of all the previous opposition parties in which Fine Gael, the successor of the pro-treaty party which had founded the State in 1922, was the dominant element.

During the election campaign some of those who now became ministers had included the steel industry among the 'white elephants' which Sean Lemass was accused of having created at public expense or risk and which it was claimed ought to be abandoned. These were, of course, electioneering speeches and did not necessarily represent political commitments. At the same time, the workers at Haulbowline were apprehensive lest a change in policy as regards the steel industry might condemn them once more to unemployment or emigration. These fears grew when it was known that other so-called 'white elephants', notably the transatlantic airline, and the hotels directly operated by the state-financed Irish Tourist Board, had been abruptly sentenced to death by the new government as part of its first economy measures on taking office.

As part of their official duties, the directors of ISH briefed their respective Ministers, Dan Morrissey in Industry and Commerce and Paddy McGilligan in Finance, about the situation and prospects at Haulbowline. It was suggested that before decisions about the steel industry were taken, the Ministers should themselves inspect the operations. Consequently on 24 April 1948 J. A. Costello, leader of the new government, Dan Morrissey and Liam Cosgrave, the parliamentary secretary for Industry and Commerce, came to Haulbowline, spent a considerable time in the works and conferred with the directors. It was widely rumoured in Cobh that the purpose of the visit was either to close the factory or to restore the original directors, whom Lemass had unseated in 1942. Neither of these expectations was realised. The Ministers were apparently impressed by what they saw and were content to allow the existing arrangements to continue, at least for the time being. That some anxiety as to policy persisted may be gathered from Lemass's appeal to the government in the Dáil when the Industry and Commerce vote was being discussed in the following July. Having summarised the history of the undertaking and its current activities he said, 'This is a key industry and should be preserved and developed as completely as possible'. In reply, Dan Morrissey did not accept or reject that view but contented himself with saying that he was awaiting a report and that the matter would be fully considered (*Dáil Debates*, vol. 112, col. 1125/26).

It was evident that the new Ministers were anxious about the possibility of having on the one hand to subsidise trading losses by ISH or on the other hand to meet the capital liabilities contingent upon collapse of the company. They may also have had doubts about the wisdom of leaving control of the company in the hands of civil servants. There is, however, no evidence that either Morrissey or McGilligan contemplated closing the steel works. The directors of ISH received the fullest support and encouragement from these Ministers. For instance, when the crisis of 1949–50 arose, as will be related, McGilligan gave Hogan carte blanche to refuse applications for foreign currency for steel of the kind manufactured at Haulbowline. At the same period Dan Morrissey was ready to ask the government to increase the duties on imported steel to the extent necessary to keep the native industry in existence. As events turned out, although it was touch and go, the directors did not ask for increased protection.

These points have been laboured in the interests of historical truth. It seems well to remove any remaining impressions that, apart from the polemics of the general election campaign of 1948, the question of continuing the steel venture was a party political issue. The political consensus regarding the state involvement in the steel industry was again illustrated when in 1954, a second Inter-Party administration came to power. William Norton, the Minister for Industry and Commerce and Gerard Sweetman, the Minister for Finance, accepted the ISH proposals for extending the range of production and for substantial expenditure on the sheet mill. By that time the company had reached solid ground. Its importance as the supplier of an essential commodity and as an employer in a depressed area had been established. At Cabinet level policy issues affecting the steel industry rarely arose except in relation to economic priorities and the allocation of resources for capital investment.

CHAPTER VI

The First Decade of State Ownership

The Company finds its feet (1947–48)

Irish Steel Holdings Ltd. was incorporated as a private company on 23 June 1947 with a token capital of one hundred shares of £1 each. Only three shares were issued, one for each of the first directors. The money to pay for the property and to keep the company in business was borrowed by the directors from the Bank of Ireland as a current account overdraft.

The directors' power to raise money was not unlimited. Under the Articles of Association prior consent of the Minister for Finance was necessary before borrowings could be undertaken. The consent was not a guarantee. The Minister could, however, guarantee repayment of ISH borrowings under an Emergency Powers Order empowering him to do so within a limit of £300,000. That figure had been proposed by the directors themselves and included the £125,000 due to the Receiver, £75,000 for repairs and improvements to the plant at Haulbowline and £100,000 for working capital. It was a modest enough financial base on which to build an important industry. Yet it proved sufficient for more than ten years.

On the day after the company's incorporation, as certified by the Registrar of Companies, the board met for the first time. Their main purpose was to approve and implement the contract made with Vincent Crowley for the purchase of Irish Steel assets. That first meeting was held in the solicitor's office at 3 College Green, Dublin and for some reason only O'Shea and Hogan were present, with Dermot McDowell who was appointed temporary secretary of the new company. An account was opened with the Bank of Ireland, the comptroller and auditor-general was appointed auditor and the shares were allotted to the directors. It was decided that Alfred Donohoe, the secretary of the former company, should be asked to remain in charge of the works until Christy Fitzpatrick could take over as the board's representative.

Before the board met again, O'Shea and Hogan had taken possession of the works and retained the key staff necessary to supervise and maintain the plant. At the second meeting, in William's office, the directors heard Fitzpatrick's report, 'The plant generally is in bad condition and the rolling mill has deteriorated due to lack of care.'

One further item in those early board records may be noticed. At the third meeting, also in Dublin, on 18 July 1947, Alfred Donohoe was appointed secretary of ISH. He reported on a wage claim from the maintenance staff and the minutes of the meeting record 'the directors' concern at the proposal to increase wages by sixpence per hour, which would mean an increase of sixteen shillings and eightpence per ton in the price of steel'. Thirty years later, after decimalisation and on the crest of inflation, one recalls nostalgically the days before the penny and the pound had lost their value.

At the outset the directors frequently met in their departmental offices, usually after normal working hours, and often without the attendance of any company official. They considered reports and recommendations from Fitzpatrick and also the serious problems which threatened the survival of the industry – difficulties in securing supplies, hostility from the scrap dealers, the pricing policy in the face of continental competition, the reluctance of the merchant stockists to place adequate orders. All three directors worked in adjacent buildings and could foregather at short notice. On the other hand it was not always easy for them to find a day on which all of them would be available to travel to Haulbowline. They were busy officials and both Williams and Hogan often had to attend meetings outside Ireland.

Thus, in the first year, apart from informal discussions, of which there were many, sixteen formal board meetings took place of which only five were held away from Dublin, either in Cork or at Haulbowline. The pattern continued in subsequent years, at least until the end of the decade. It was not altogether a good practice. The directors were really acting as a 'committee of management' rather than as a policy-making board and they were dealing with too many details of administration. Williams, indeed, was operating virtually as a managing director, particularly in times of crisis. As a result the length of agenda and duration of meetings were far greater than would normally have been acceptable to part-time directors or conducive to the growth of strong and efficient management. One unhappy person in this environment was the secretary, who felt obliged to resign in November 1947. The vacancy was filled by another qualified accountant, Alfred Tutty.

Doubtless the directors' continued intrusion into management was prompted as much by their determination to make a success of the project as it was by anxiety to avoid or neutralise mistakes by inexperienced management which could cause financial damage or prove politically embarrassing to Ministers. There was also the fact that because of their official power, the directors were able to deal more rapidly than management could with some matters affecting the industry. For instance Williams brought pressure to bear on the Board of Trade to give export licences for supplies of coal and silica bricks urgently needed for continued production at Haulbowline, and Hogan 'as a special case' authorised provision of hard currency to purchase ferro-manganese which could not otherwise be procured immediately. This wearing of two hats by civil servants was, of course, anomalous and there were some who complained that the directors were bending official policy to suit the needs of the steel industry. Arising out of one notable and persistent complaint by A. P. Reynolds, chairman of the Irish Railways (CIE), the directors found it necessary to establish by board resolution, to be sent to the Minister, 'that their actions as directors in the interests of the company were entirely independent of their functions as civil servants' (August 1948).

When the board met for the first time at Haulbowline on 1 November 1947, they were able to consider progress since Fitzpatrick had succeeded in commencing commercial production six weeks earlier. Ingots had been produced at £25 per ton, billets at £32, bars (single rolled) at £34 and (double rolled) at £41. The average selling price, including extras, was £39 per ton approximately. Owing to lack of materials, mainly scrap, production had been restricted to supplies for essential manufacturing industries.

The board's policy at this juncture was to concentrate on production from the existing plant so as to meet as much as possible of the home trade requirements. There is a significant entry in the minutes of 28 February 1948, when Fitzpatrick was told to 'suspend all investigations regarding the suggested extension of the industry and carry out only such maintenance work as is necessary to keep the existing plant in production'.

The portentous visitation by the Taoiseach and the Minister for Industry and Commerce in April 1948 has been mentioned earlier. In the following October the same Minister, having seen the company's first annual report, asked that his appreciation be conveyed to all employees. Perhaps foolishly, the board directed that a note about the Minister's message be inserted in the next pay packets. The predictable response was a demand for increased pay by the workers, leading to a long dispute during which there occurred the summary dismissal of all the foreign furnace men in December 1948, and the prosecution of one of them for malicious damage to equipment. This was a turning point in the industry. Fitzpatrick's confidence that he could rely on 'green' Irish labour to replace the foreign operators was fully justified following an intensive spell of training. It was still necessary, however, to rely on imported skills for two supervisory posts in the plant. The rolling mills superintendent was P. Austin, a product of the British steel industry who had come to Irish Steel Ltd. in May 1946 on retirement from a Turkish company. W. Rowlands, the melting shop foreman, had been recruited in England by ISH in July 1948 to replace J. Davies (mentioned in chapter four), who resigned after a year with the new company. Both Austin and Rowlands were practical men of long experience; they served their new masters loyally and trained good Irish workers to support them. Bill Rowlands resigned in October 1956, being succeeded by his assistant, John Rafferty, a local man of exceptional quality. Austin died six months later. It was not easy to replace his experience at once. While solutions were being found, much of the credit for the continued efficiency of the mills could be attributed to George Radley and also to R. H. Warbrook, an engineer and manager recruited from the Mallow Sugar Factory in January 1956.

At the first annual general meeting on 9 October 1948, the financial results of the accounting period ended on the previous 30 June were tabled. A loss of about £500 was reported, not a bad result considering that the company had been trading for only part of the period. Williams's statement to the shareholders, in effect the Minister for Finance, whose permanent secretary, J. J. McElligott, attended, was given in low key. The main achievement had been the steady employment at the end of the year of a labour force of over 300 with ability to produce steel of good quality. An independent valuation in June 1948 had shown 'the value of the premises and plant to be much in excess of the figures in the Balance Sheet'.

Crisis and survival (1949–51)

During 1947 and 1948 there had been little to worry about in the marketing field. At a basis price of £38 per ton (plus extras) Haulbowline steel was competitive with duty-paid British steel. The problem was to produce enough Irish steel, production being inhibited by shortage of materials, particularly scrap, labour difficulties and some plant deficiencies and breakdowns. In mid-1949, a tendency emerged for continental steel to come on offer at low rates. The German steel industry and also plants in Belgium and Holland, which had been devastated during the war, were resurrected dramatically with American aid. As a result outputs of steel exceeded demand on a world-wide basis. The market began to be flooded by offers at what were really dumping prices.

The directors of ISH reacted to the situation in June 1949 by reducing the basis price by £3 per ton and the size extras by 30 per cent. By September the situation had worsened due to a lack of orders and quite clearly the company had no financial resources to enable it to continue in production except against firm orders. A series of meetings took place between the directors and the Minister for Industry and Commerce and subsequently with the Industrial Development Authority (IDA) and the National Association of Iron and Steel Merchants.

The steel merchants and steel stockholders held a key position. They were fearful that, if the downward trend in foreign steel prices continued, they would incur heavy losses by buying Irish steel for stock and consequently they would not place orders without assurance that the government would take action to keep the foreign steel out, or else subsidise Irish steel prices so as to make them competitive. Such last resort measures as increased tarriffs, or import quotas, or subsidies, were deferred by Ministers while the directors renewed their own efforts to avoid closure of the works. The devaluation of sterling, and thus of the Irish currency, in mid-September 1949, made steel imports from the Continent dearer. This benefit was, however, rapidly eroded when continental export prices, notably from Belgium, fell lower still. The exchange control regulations limited or delayed foreign payments by Irish merchants but these restrictions could help ISH only so far as the merchants would buy instead from Haulbowline.

In the directors' view, the merchants were unduly apprehensive and some of them were hostile to the native industry. To allay the merchants' fears, they offered special prices for large orders and they promised to give a rebate on stocks held by merchants at the date of any future price reduction. They circulated government departments and local authorities asking them to insist on the use of Irish steel in their contracts. They decided to offer Irish steel at sacrifice prices for large orders to be exported abroad, so as to maintain production and employment.

Notwithstanding all these efforts, the situation worsened and on 29 May 1950 the directors, meeting in Dublin, discussed 'the precarious situation which has arisen for the company'. They decided that to be competitive they must produce double-rolled sections from imported billets which could be bought cheaply. Within limits they would roll for stock rather than lay off men. If the merchants would not buy, they would break with the Association and sell at their basis price to any buyer. The basis price was intended only for approved stockists and manufacturers who placed orders regularly for minimum tonnages fixed by ISH.

A month later at a joint meeting with the IDA and the merchants, the directors made their last bid for orders. They reduced the basis price to £32 and offered to sell at £30 per ton for large orders placed immediately i.e. within fourteen days. Not a single order was placed at the meeting and the directors came away despondent. They were aware that reinforcing bars were on offer that day f.o.b. (free on board) Antwerp at £18/£19 per ton and could be brought into Dublin, duty paid, for less than £30. However, some merchants did place orders and production at Haulbowline continued.

Events on the other side of the world came to the rescue of the hard-pressed industry at Haulbowline. The invasion of South Korea by Communist forces at the end of June at once created pressure on steel supplies for re-armament. On 12 July 1950, the directors met the Ministers for Finance and Industry and Commerce, and measures were agreed for the maintenance of production and employment and the purchase of imported billets so that stock-piling of steel for emergency conditions could take place. Even though the battle-field was confined to Korea, where the United Nations forces were confronted by the Chinese Communist armies for more than two years, the danger of a much wider conflict turned the world steel market upside down. As early as 15 August 1950 substantial increases in the offer prices for British and continental steel delivered to Ireland were reported. Not long afterwards these prices had escalated to such a degree that Irish steel became the cheapest in the world. The directors, satisified to have regained their home market through the chances of war, resisted the temptation to increase prices forthwith beyond the level justified by costs and reasonable profits. At the third annual general meeting of the company on 24 November 1950, Williams made the position plain 'it is the policy of the directors to keep Irish steel prices as low as possible'. He also stressed the importance of giving priority to the needs of home customers, even when the steel could be sold to other countries at higher prices due to temporary circumstances.

The sellers' market for steel caused by the Korean war was naturally extended to steel-making materials and to semi-finished products such as billets. More had to be paid for

Irish scrap if supplies were to be maintained. During 1951 ISH selling prices rose on account of costs and in August of that year the basis price per ton stood at £40/£45 compared with the £32 fixed in the dark days a year earlier when the order-book was blank. The new prices were still below any alternative open to an Irish buyer, if, indeed, supplies from abroad could be secured at all without long delays.

The marketing crisis was succeeded in mid-1951 by a production crisis. Due to a virtual boycott by some of the scrap dealers and to the policies of CIE (*Córas Iompair Éireann* – Irish Transport Company), who wished to export railway scrap, ISH had never succeeded in establishing an adequate and timely flow of scrap to Haulbowline. It was mainly due to the good offices of the leading scrap holder, Hammond Lane, that a scrap famine had been averted in the early days of the steel company. Now the situation had become so serious that Williams, putting on his hat as chairman, wrote on 20 July 1951 in plaintive terms to the Department of Industry and Commerce (of which he himself was second in command). This letter mentioned twelve manufacturing firms who would have to cease operations if ISH failed to supply steel: 'It is almost impossible to obtain steel elsewhere although Belgium will supply slowly but at prices about twice that of Irish steel'. There were only one hundred tons of scrap in stock and five hundred in sight. CIE was blamed as the principal defaulter in supplies, from which source 10,000 tons of good quality scrap normally came each year. Other suppliers had stated that they had none or that it had already been sold. 'I must state', ended Williams, 'that at the moment the directors cannot see what the future holds for the steel industry or those dependent on its output.' A few days later the board formally requested the Minister to enforce rigidly the prohibition against scrap exports and to announce his intention of doing so. Following meetings with CIE, the situation eased, ISH for its part offering premium payments for large deliveries of scrap into the steel works. Thus another crisis dissolved, but the problems remained.

During the period of crisis the company had two good strokes of fortune. Tom O'Shea, who had retired from the civil service on pension the previous year, agreed to give 'his full time to general supervision of the undertaking for a period pending the appointment of a new secretary'. This was the first of many assignments undertaken by O'Shea, who had time to spare, mainly in the field of marketing and public relations but also in chairing works committees on development. For a dozen years he was active in these spheres, much in the role of an executive director without the title or pay and, in particular, he did great service in promoting goodwill between the company and its customers. The second fortunate event was that Gerard Farren joined ISH as secretary/accountant in August 1949, being the third to fill that position since the company's inception. His predecessor A. J. Tutty had gone to a job abroad. Farren, a certified accountant with commercial experience, including service with the Dublin port authority, had the ability, dedication and stamina which are essential to management. When he presented the financial accounts for the company's second year, 1948–49, they showed a net profit of £20,000, and continuous production during the year by a labour force of 340.

Farren had a less pleasant duty on 17 November 1949: he reported to the board the first fatal accident in the industry, the unfortunate P. Snow having been killed during unloading of a barge at the dock-side.

Consolidation and re-assessment (1951–53)

When the three civil servants were instructed in April 1947 to purchase the assets of Irish Steel Ltd. and to revive the industry they also had a third task to perform. As its name implied, the company formed to conduct the new venture was conceived as a temporary or 'holding' arrangement. Its directors were expected, after some experience of the steel industry, to furnish to the government a report on the situation with recommendations as to what steps might be taken, either by the state or by private enterprise, to ensure a viable future for the undertaking. In February 1949 the directors had sent a twenty-five page report to the Minister for Industry and Commerce, which he circulated to the government. This document, although signed by all three directors, was unmistakably the work of Williams, whose succinct and vivid style was rare among civil servants.

The report described the Haulbowline works and the steel-making and rolling processes. With expert advice obtained by the directors after they took office, many repairs and additions to plant and equipment had been effected. The difficulties experienced in securing essential materials were described as well as the problems of recruitment, training and remuneration. On the basis of these explanatory details the report appears to adopt two stances, one self-congratulatory in the sense of 'mission accomplished' and the other pessimistic as regards the long-term prospects for the industry when faced with keen competition from foreign steel. The directors admitted the insufficiency of the evidence. 'Our experience', they wrote, 'of this important industry is relatively brief and it is difficult for us to make a confident recommendation. In the course of a few weeks one swings from the highest hopes to the depths of despair and back again. In any event, we have demonstrated, and for the first time in this country, that first-class steel can be produced here on a continuous process solely from native scrap. Irish labour is employed on 90 per cent of the processes and in time should be able to take over practically 100 per cent. The works and plant have been improved beyond anything that we thought possible.' The directors did not favour embarking on the erection of the blooming mill or the sheet mill and considered that further capital should be spent only on the improvement of existing equipment and handling facilities (estimated at £50,000). They saw no prospect whatever of any investment by private enterprise.

The report concluded that on account of the itemised cost disadvantages (at 1947 prices) as compared with British steel the existing industry could survive only with the assistance of high tariffs or subsidies. It might be advisable to consider the abandonment of steel-making at Haulbowline and setting up an alternative industry by moving the rolling mills to a Dublin site where a viable future might be found in rolling imported billets into a wide range of bars and sections required by the Irish market. This solution presumed that billets could always be bought at economic prices. The directors did not find that idea very attractive but 'it might be considered in the light of further experience of working the present plant at Haulbowline'.

This somewhat inconclusive report was referred by the Minister to the IDA (Industrial Development Authority) for their opinion. The crises described in the preceding section of this chapter pushed the report into the background and there is no record of any subsequent decision upon it by either the Minister or the government. The report was, however, still relevant in June 1951 when, with a change of government, Sean Lemass once again became Minister for Industry and Commerce and turned his attention to the steel industry.

At a meeting with the Minister in July 1951, the directors were informed that the government intended to maintain the steel industry and would, if necessary, introduce financial legislation. At this meeting and subsequently at Haulbowline on 12 September, his first visit to the steelworks since the day twelve years earlier on which he had officially opened Davy Frame's rolling mills, the Minister asked the directors to report on the possibilities of expanding the range and output of the industry. Lemass felt that reasonable risks should be incurred in building on what had already been achieved.

The directors were, themselves, by no means complacent. During 1952 and for most of 1953 the demand for steel was spasmodic and at times the recession was serious. The irregularity of scrap supplies continued to be a problem which the directors again sought to solve by increasing the basic price by ten shillings per ton and also offering an 'inducement bonus' of nearly a pound per ton for regular large deliveries. The latter device which remained an essential cornerstone of the ISH scrap purchasing system was challenged before the Restrictive Practices Commission twenty years afterwards by people who alleged that the bonus arrangements unduly favoured Hammond Lane Metal Company and unfairly eliminated small scrap dealers.

As an earnest of the the government's support for the steel industry the Minister for Finance (McEntee) made a reference to it in his Budget of April 1952. He stated that while hitherto Irish Steel Holdings had been financed by bank overdraft on both capital and current accounts, it was undesirable that this should continue. A sum of £250,000 was included for the steel industry in the Capital Budget for 1952–53. As it turned out, this subvention was not paid because the company's cash flow was healthy and it did not need the capital pending completion of plans for development. The government's willingness to invest was a help to the directors in their negotiations with the bank for continued overdraft facilities. A less helpful sequel was Deputy Corry's persistent queries afterwards in Dáil Éireann as to 'what had Irish Steel done with the quarter million the government gave them to erect the sheet mill' and nothing would persuade the Deputy that money included in the Budget and approved by the Dáil had not, in fact, been paid.

In November 1952, John Williams reported to the fifth annual general meeting on the results of the company's financial year 1951–52. Turnover was £865,000 as compared with £640,000 in the previous year and £428,000 in 1949–50. Net profit after taxation was £66,000, compared with £55,000 and £39,000 in the two previous years. Employment was 450, which was nearly double the number engaged when the industry began operations five years earlier. Williams was guarded in his comments about the future as he had every reason to be in view of the depression in the home trade and the difficulties with supply of essential materials, particularly scrap. While 1952 drew to its end, Fitzpatrick reported 'No orders' and the directors decided to roll up to 1,000 tons for stock. Continental prices fell and there was anxiety as to the adequacy of the existing protective tariffs. ISH prices were still below those available from Britain where, moreover, supplies were scarce.

The first moves towards expansion (1953–57)

During the year 1952, the directors continued their practice of conducting formal board business mainly in Dublin. Of twenty-two board meetings during that year only four were held in Haulbowline. The directors visited the works on other occasions. The chairman was there frequently as was O'Shea who kept an eye on sales policy and customer relations.

The year 1953 opened badly. The recession in home demand persisted. Export attempts, even at the sacrifice prices offered in Britain and Northern Ireland, failed. In June a small order of 330 tons of bars to be consigned via Newport to

Portugal was hailed as a breakthrough. There was no shipment, however, because the customer wanted too much credit.

It was against this somewhat gloomy background that the directors, in February 1953, submitted to the Minister for Industry and Commerce a fifty-page report on 'the present position and prospects for development'. While satisfactory results had been obtained from the existing plant, and tributes were paid to staff and workers for their achievements, the report took a conservative view of the possibility of increasing outputs beyond the then level without installing additional equipment. It was also pessimistic regarding probable future growth in home demand for steel products. However, some of these assumptions proved false within a few years when the two furnaces were operated in tandem for long spells and when producer gas was abandoned in favour of fuel oil. Forecasts of Irish scrap supplies and of product demand proved too low.

The argument in the report centred on development and, especially, on the possibility of successful sheet production at Haulbowline. The directors gave warning of the danger of optimistic assumptions based on the recent success of the industry. 'Great care', they wrote, 'must be exercised to ensure that efforts towards development are such as will not jeopardise the financial and economic structure of the industry on its present basis.' They also felt that any large-scale deployment of foreign technicians and operatives in order to achieve rapid development would be a mistake. Development in their view should proceed gradually on the basis of the industry growing its own technical skills and worker proficiency.

The directors came to the conclusion that, but for the fact that the existing incomplete mill had cost the company virtually nothing and could be put in operational condition for about £120,000 'no one could seriously contemplate the manufacture of sheet steel in this country as a sound industrial venture'. The report gave a guarded opinion in favour of completing the mill so that it could be disposed of abroad, even if production at Haulbowline were not attempted. If the attempt was to be made, the report was positive that this step should be preceded by installing a pickling, galvanising and corrugating plant at Haulbowline, based in the first instance on imported sheets. This was so both for marketing potential and in order to give the workers experience in handling sheet metal.

In the result, the only positive recommendation for development in the report was adopted by the government in August 1953. The International Construction Company of London was appointed consultants to the project. O'Shea, at the request of the board, undertook to attend at Haulbowline on as many days as possible for three months 'in a supervisory capacity'. In due course, after competition, contracts for the equipment and for its installation were placed with Taylors of Briton Ferry for the pickling plant, and with Eichener of Germany for the galvanising line and corrugating press. These jobs were eventually completed for about £70,000.

In December, Industry and Commerce conveyed approval in principle for the sheet mill project and an intention to promote legislation to capitalise the venture assuming that on further consideration the directors would put forward firm financial proposals. These were submitted in May 1954.

Shortly before the government approval, Williams had presided for the last time at a general meeting of the company, when he reported for 1952–53 a reduced net profit of £22,000 due to fall in home demand, increased material costs and a six weeks' closure of the works caused by an unofficial strike of the electricians. Shortly after this meeting he became ill and died in March 1954. John Williams had for seven years given leadership and purpose to the difficult task he shouldered on top of his manifold other official duties. The board minute, recording with sadness his untimely death, rightly referred to him as 'the foundation stone and inspiration of the industry'.

In June 1954 a change of government took place bringing to office an Inter-Party administration. William Norton, leader of the Labour Party, became Minister for Industry and Commerce and also Tánaiste (deputy prime minister). On 18 June he received the two directors and expressed his approval of the industrial operations at Haulbowline and his intention to support them. Later on the Minster visited the works when the galvanising line was in operation and encouraged the company to proceed to further development.

During 1954 trade improved considerably. Fitzpatrick

pressed forward the work of installing the galvanising plant and also the new reheating furnace, ordered from OFU in Germany at a cost of £40,000. The company's own resources were adequate to finance these capital provisions without recourse to government funds. In October 1954, however, it was found necessary to persuade the Bank of Ireland to increase overdraft facilities to a level of £400,000 without any additional state guarantee.

In the midst of his preoccupations with production and development, Fitzpatrick was faced in August 1954 by a 'walk-out' of the workers on the merchant mill, who objected to his rolling programme. After consultation with his acting-chairman, Hogan, the general manager took a hard line. He declared that the unofficial strikers would not be re-employed. Then he proceeded to work the merchant mill under his direct personal supervision, using non-unionised mill personnel reinforced by gallant volunteers from the office staff. The strikers at Cobh were expected to be startled and chagrined by the clang of the rolling mills which could be clearly heard across the water. There was no sympathetic strike action by other steel workers. After ten days, at the request of the Union, the company took back most of the strikers, but not the ring-leaders, who forfeited their jobs. Fitzpatrick had proved his point. This incident is memorable and so recorded only because in the light of present-day management/labour attitudes, it seems incredible that it ever could have happened.

The seventh annual general meeting of the company was held on 12 November 1954 with Hogan in the chair and Redmond, secretary of the Department of Finance, as proxy for his Minister in attendance. A net profit of £61,000 for 1953–54 was reported. Selling prices had twice been reduced during the year in keeping with the policy of giving customers the maximum advantage from the company's growing output and efficiency. Under the Finance Act of 1954, the company became exempt from corporation profits tax because its Articles of Association precluded distribution of dividends. In that context the chairman mentioned that total tax payments by ISH since its inception had been £120,000, not far short of the fixed capital base on which the company had been established, namely the purchase price of the assets.

For a while the galvanising plant and the sheet mill project occupied the front of the stage. In both of these activities, a new acquaintance emerged in the shape of Richard Thomas and Baldwins (RTB), the well-known British steel company which had been nationalised in 1949. Fitzpatrick had established contact with RTB through its managing director, H. F. Spencer, in May 1954 and those relations were confirmed when the directors visited London in June. As a result, training for ISH operatives was arranged in one of the RTB South Wales plants and assurances were received regarding the supply of black sheets (then in short supply) for Haulbowline. The Minister for Industry and Commerce secured a government Order imposing tariff protection for the new ISH product at rates parallel to the existing *ad valorem* duties on steel. The works' records show that the new plant, after some delays, was producing commercially in November. In January 1955 Fitzpatrick reported to his board that output levels were in line with those achieved in similar shops in South Wales. Market demand was so favourable that a double shift operation was begun with additional employment. While in the early stages some quality complaints were received, causing headaches for O'Shea as director in charge of sales, experience both by the producers and by customers in proper storage of the sheets brought satisfaction. In the years ahead the galvanising operations often proved a financial boon when other Haulbowline activities were in a depressed state. The demand for corrugated galvanised sheets was greatly stimulated by the government grants for farm buildings instituted in 1956 with conditions imposed regarding the use of Irish materials.

One of the board's worries concerning the sheet mill venture was the danger that, if it failed, it might bring down the whole industry. This aspect had been touched on in the report to the Minister. Some years previously, Hogan had discussed the possibility of a joint venture with H. Benedict, a vice-president of Armco. Subsequently a representative of the American firm inspected the plant at Haulbowline. Nothing came of this, nor of some polite concern shown by Metropolitan Vickers and by Demag. However, RTB, as potential suppliers of sheet bars for rolling at Haulbowline, was in a position to have a closer interest. In process of rationalisation they were closing their own small sheet-rolling and galvanising shops for which they produced sheet bars, and concentrating on a larger modern galvanising unit associated with their new continuous strip mills. Having first cleared the point of policy with Ministers Norton and Sweetman, the directors and general manager aired a proposition in London on 8 February 1955 over lunch with Sir Ernest Lever, chairman of RTB, H. F. Spencer, and Bill Cave their sales director. To add weight to his team, Hogan asked F. H. Boland, the Irish Ambassador, to join the discussion. The ISH ideas were based on the creation of an Irish company to produce sheet at Haulbowline capitalised in equal shares between the two partners, the operation to be structurally and financially independent of ISH, although closely related on the same site. The matter was pursued at a further meeting with the same RTB representatives in London in the following August and progressed after that meeting to the point that in November Hogan and Roberts, the RTB financial controller, prepared heads of agreement for a joint investment of £500,000. However, not long afterwards, RTB drew off, as they said on account of the opposition of their Minister of Power and the Iron and Steel Board who considered that all of the nationalised company's financial resources should be devoted to the great new strip mill being planned for Newport. The outcome had some similarity with that which occurred a dozen years later in another case of promising negotiations. In both instances, the British government attitude was a factor.

Following the death of John Williams, the board of ISH was in the hands of the two surviving directors until March 1955 when J. C. B. MacCarthy was appointed chairman. His administrative career had been in Finance, Supplies and Industry and Commerce and eventually as chairman of the Restrictive Practices Commission from which post he was recalled to Industry and Commerce in the reorganisation which followed Williams's death. A man of rare abilities and a most skilled negotiator, he brought a fresh mind to the problems of Haulbowline.

In June 1955, Fitzpatrick gave notice of his intention to resign his post at the end of July. The board asked him to remain with the company in a temporary capacity while reorganisation was effected, and he agreed to stay until the end of the year. During this interval, Farren, the company secretary, who had previously acted for Fitzpatrick during the latter's illness, was designated manager and efforts were made to find someone with experience suitable for a post of works manager in a steel industry, duties which Fitzpatrick had included in his role as general manager. In the event the decision was that Farren should succeed to the full post of general manager and that in the control of the shop floor, he should be assisted by a works superintendent, R. H. Warbrook.

While so much occupied with personnel problems and indeed also with wage claims, MacCarthy still found time to continue negotiations with RTB and to press ahead generally with the sheet mill project as desired by the government. In September 1955, accompanied by O'Shea and Radley, he visited Germany, inspecting sheet mill plants and exploring commercial prospects. The purchase of materials on a regular basis was a continuous problem for ISH. The policy followed was to have plurality in suppliers and not to rely on a single source. In November Hogan and Farren went to London to conclude firm engagements with the National Coal Board and also to re-open the possibilities of RTB involvement in sheet steel production.

During 1955 and the first half of 1956, production and sales generally increased. The steps Farren took to operate the open-hearth furnaces in tandem for longer spells gave increased ingot output, although it was still necessary to import billets to maintain the merchant mill's full potential. MacCarthy was able to report favourable results for 1954–55 at the annual general meeting in November, the accounts showing a net profit of £61,000. Redmond again attended as proxy and joined in the tributes to Fitzpatrick 'for work he has done so ably since the inception of the company'.

It had been one of the boasts, if that is the correct term, of management that alone among steel works ISH furnaces produced steel solely from native scrap. The iron necessary to supplement steel scrap in a satisfactory charge was furnished by cast iron scrap. As ingot output increased it was, however, necessary to import pig iron. The early 1956 records show imports of German pig iron at £38 per ton, a high price. At this stage, ten years after the war, steel prices and raw material prices were still unsettled. ISH found difficulty in buying billets and black sheet on a satisfactory basis as to both quantity and price.

In June 1956 Hogan succeeded MacCarthy as chairman of the board, when the latter became secretary of the Department of Industry and Commerce. MacCarthy remained an ordinary director until August when he resigned as it became impossible for him to attend meetings. From that time until the following December, Hogan and O'Shea once again found themselves the sole directors. William Norton discussed with the chairman the merits of appointing a Cork man, preferably one with engineering or other technical attributes. However, eventually, the Minister's choice fell on John Connor, assistant secretary in Industry and Commerce, thus preserving for a further period the civil service dimension in ISH. Connor played an important part in the activities of ISH for ten years, his very definite views and pungent comments adding flavour to board meetings.

The second half of 1956 brought a recession in home demand for both bars and galvanised sheet which lasted for nearly a year. There was a general international slump, affecting Irish economic conditions, with reduced constructional activity and unemployment. The directors took measures to reduce expenditure, curtail imports, run down stocks and sell surplus stores. As assistance to the national economy ISH granted a special price discount to manufacturing firms using Irish steel in their exported commodities. For further efficiency and cost reduction, the board decided to convert the furnaces to oil-firing, thus making the gas producer plant surplus to requirements and available for re-sale or as scrap. In order to maintain employment intense efforts were concentrated on finding export outlets, with some success as is related later. For the same employment reasons, the board directed in April 1957 that special works of maintenance and road-making should be undertaken.

These recessional problems were accompanied by bank pressure to reduce overdraft facilities. Early in 1957 a twenty-five year loan of £150,000 at 5½ per cent interest was obtained from the Irish Life Assurance Company. This loan was state-guaranteed, the Bank of Ireland having agreed to reduce its own guarantee from £300,000 to £150,000. At the annual general meeting in November 1956, the chairman had been able to report that works output had risen by 40 per cent and that record sales of £1.5 million had been achieved during the year. The net profit was £73,500. A year later when the tenth annual general meeting was held, the accounts for the year 1955–1956 showed a fall in profitability to £55,000 but this was reported as a favourable result in view of the recession in home demand and of the intense competition in export markets. The increase in ISH selling prices had been moderate. Over the past two years Irish prices had risen by 11 per cent for bars and 4½ per cent for galvanised sheets as compared with British price increases for the same products of 21 per cent and 17 per cent.

At that meeting in November 1957 the chairman's statement also surveyed briefly the achievements of the company during its first decade. Production had been quadrupled, wage disbursements had been trebled, sales totalled £8 million with accumulated net profits of £450,000. The Exchequer had benefited by tax payments of £200,000 and, contrary to much public misconception, the industry had not received a penny in state subsidy or capital subscription.

The labour force and industrial relations

The pages of this book, perhaps inevitably, seem to recount mainly a series of events springing from the motives and decisions made at certain stages by a few individuals or else the result of policies and objectives formed either in the corridors of government or around the board table. None the less the steel industry, as a collective force and a step forward in economic progress, depended for its beginnings and its continuance on the unnamed men who manned the furnaces and the mills, operated the cranes and works transport, or maintained the power, heating and mechanical installations in good working order. There was no steel-making tradition around Cork harbour or anywhere else in Ireland. During the time of the first undertaking, Irish Steel Ltd., the men were more often idle than they were employed in production. The intermittent activities had been in bar rolling operations until May 1946 when the two Lambot open-hearth furnaces were commissioned. Furnace operation then lasted for a few months. Both on the furnace floor and in the mills, skilled positions were filled by foreign operatives who supervised the Irish semi-skilled and unskilled labour. On account of the spasmodic production, training and efficiency must have suffered. That the Irish workers were adaptable and proficient in acquiring new skills was proved by subsequent experience when continuous production became possible under the new company.

Thirty years afterwards, there were still with ISH a few survivors of the gallant pioneers whose efforts and example made possible all that followed. Their names may well be mentioned, together with their grades in December 1976: J. Rafferty, asst. general manager, G. Radley, mills manager, D. Lynch, asst. manager (mills), F. Leonard, training officer, T. Cronin, maintenance engineer (mills), J. Hallorhan, maintenance superintendent, J. Mahony, supt. construction shop, M. Ryan, executive, galv. shop, W. Downing, sample passer (steelworks), M. Harley, roller (430 mill), P. Riordan, roller (430 mill), R. Hally, despatch clerk, W. O'Riordan, asst. despatcher, P. Mahony, dockside foreman, C. Gleeson, foreman bricklayer, W. Drake, foreman (stock control), C. Pritchard, foreman (electrics), J. Owley, foreman (galvanising), J. Barry, mate (machine shop), M. Desmond, first aid attendant, F. Devlin, asst. roller (430 mill), J. Kidney, chargehand electrician, S. McNamara, mate (roll turning), J. O'Sullivan, combustion (430 mill), S. O'Sullivan, boiler house operator, G. Quaine, plumber, M. Sliney, chargehand roll turner, T. Twomey, roll turner.

ISH inherited both the union and the wage structure already in existence at Haulbowline. The bulk of hourly-paid unskilled or semi-skilled workers were members of either the Irish Transport and General Workers Union or the Workers Union of Ireland. Skilled workers, many of them foreign, and foremen or their equivalents were for the most part non-unionised as were supervisory and technical grades. The various tradesmen such as electricians, fitters, plumbers, roll-turners, carpenters, etc., were members of established craft unions, some of them British based. The administrative and clerical staffs were non-unionised.

At an early stage in ISH history, for reasons which are not recorded, the Workers Union of Ireland ceased to have members employed in the steel works, so that the ITGWU became the sole union for hourly-paid unskilled or semi-skilled workers, who formed the large majority of those employed by ISH. As early as August 1948, the company entered into an agreement with that union recognising its paramount position in relation to all the manual workers exclusive of craft workers and supervisory grades. This dominance of the ITGWU was gradually pushed by the union to a position of 'exclusivity' and finally to the reluctant acceptance by the directors in 1957 of a closed shop; men recruited by the company for the mutually agreed grades of work, if not already members of the union, must join it within a limited period or else be liable to dismissal. The reluctance was not caused by distrust of the union concerned, with whom relations had been good, but was due to the constitutional and legal hazards the company might face if the union pressed the point against a man already employed who did not wish to join any union or contribute to its funds and sought the protection of the courts.

Indeed, on many grounds the company would have preferred to deal with a single well-organised union. It was not possible, owing to the fiercely independent attitudes of the craft unions, of which there were more than one for some of the trades, to establish a situation such as had developed in some foreign countries. In these economies all workers throughout the steel industries, whatever their duties, were grouped in and represented by a single trade union formed for that purpose. This was unacceptable in Irish conditions. The plurality of unions at Haulbowline led to leap-frogging in wage claims, and difficulties for management in negotiating wage structures in keeping with the economics of the industry. The key position of the craft workers, such as electricians and fitters, whose maintenance and repair duties were essential for continued production, put them in a strong position in negotiation. The ITGWU and its members in the industry were, naturally, watchful and resentful of any concessions to the craft workers which they considered should be applicable also to other employees, even though the basis of differentials could be defended by management and was in keeping with practice in other industries. The tug-of-war at Haulbowline between the craft and non-craft unions has always been one of the trials of personnel administration in ISH.

In 1957 at much the same time as the ITGWU were seeking a closed shop in its own sphere, and indeed that move may have been influenced by this next event, the Irish Union of Distributive Workers and Clerks began to organise the hitherto non-unionised clerical workers in ISH. This was by no means resisted by the company, except in one aspect. It was felt by management, with the full support of the board, that staff officers, superintendents and female confidential secretaries to senior executives should not join the union. The officers concerned accepted this restriction but the union fought the matter for some time, eventually agreeing to exemption of specific named persons and not to the principle of exclusion of grades normally regarded as clerical and not managerial. All this debate has an old-fashined look. It is no longer regarded as either politic or feasible in industrial relations to prevent any employee, whatever his or her rank, from joining a union which has been accepted as a negotiating body by the employer involved.

When ISH first employed industrial workers in 1947 the rates of pay were conditioned by the practice of the previous company. The basic pay for an unskilled worker was related to the local builders' labourers' rate with additions for semi-skilled employments. Craft workers were paid at the rates in being for similar trades in the Cork area. Clerical rates in a locality of high unemployment were very much ad hoc and would seem to have been at a low level. For instance in 1949, the commencing pay of a typist was £2 per week; a girl who could take shorthand received an additional twelve shillings.

Claims for increased pay for the factory workers were pressed by the unions almost from the outset and equally strongly resisted by directors intent on giving the industry a chance to survive its infancy. In January 1949 both parties agreed to submission of the current claim to the Labour Court, as a result of which the basis of remuneration for hourly-paid workers was established that in essentials was maintained throughout the company's first decade. A basic hourly rate, reflecting appropriate local comparable rates, was supplemented by a production bonus geared to the outputs of the furnaces and of the rolling mills. The better the weekly performance, the larger the pay packet. When tonnage outputs were high, the bonus could represent as much as one-third of the total pay. In times of poor production or of plant failure, wages fell and consequently there was always a plea for an increase in the stable element of the pay. Shift workers received supplementary payments.

In general, workers on the furnaces and their ancillary equipment received higher total pay than the men in the mills; the craft workers had differentials bringing them above other grades. It was a complex and, in some respects, illogical system even at its birth. Further refinements and additions over the years, in response to claims and the needs of equity, made it an instrument of torture for management and wage clerks. The central idea of the tonnage bonus was a good one, as it rewarded effort and performance that was of benefit to the company's financial results. However, being of universal application to the whole group of workers included in the system, it was not elastic enough to enable special premiums to be earned by exceptional individual performances, a method of reward which was, in any event, frowned on by the unions.

From the outset recruitment, wage claims and relations with the unions were handled by the general manager personally, assisted by the secretariat. This added much to the general manager's burdens. It was not until September 1961 that a personnel officer of experience was recruited, J. Nolan, who previously served in similar tasks with CIE. The directors, as their minutes show, encouraged management to take a firm line with unreasonable demands or threats of strike action and particularly in cases of unofficial stoppage of work. The chairman sometimes was required to intervene personally in discussions with union representatives, usually to make the point that if the works remained closed for any length of time on account of strike action, it might never re-open. The state, the chairman said, would not support an insolvent industry in order to give employment. If the workers thought otherwise they were gravely mistaken.

Looking at the panorama of the first decade of ISH, it seems to have been an era of industrial peace. Strikes were rare events and only in one instance, that of the electricians in November/December 1952, does the stoppage seem to have been official. It brought the industry to a standstill and was settled at a conference, over which the Lord Mayor of

Cork presided, by a wage increase of general application to all workers.

Unofficial stoppages were, on most occasions, caused by reactions of individual workers against works' conditions and disciplines. Support for the company in the face of unofficial strikes was usually given by the unions and, in this respect, the understanding and co-operation of P. J. O'Brien, the Cork area secretary of the ITGWU, was a notable factor in the very good industrial relations enjoyed by ISH during the decade after 1947. Both O'Brien and his Cobh Branch secretary, Jerry Kelly, understood the difficulties of the steel industry and the threats to its survival as did the great majority of their members at Haulbowline. As they had done in August 1954 when the mill crews walked out, so also in June 1957 the union officials fully supported the company when it dismissed two crane drivers for disciplinary offences. The craft unions tended to be more difficult and militant, but they too were not disposed to condone unofficial strike action.

The directors usually did what they could, in the context of the company's financial position, to meet reasonable wage demands, as when a further general increase was conceded in July 1955. At this period basic wages could not be regarded as high. The average for all ISH hourly-paid unskilled workers was about £6 per week. The company was anxious to appear a good employer within its means. In 1951 a small ex-gratia Christmas bonus of £3 for all workers was instituted. This was abandoned in 1962 when the wage structure had greatly increased and the company was losing money. Canteen facilities for employees were established in due course and subventions by the company towards a benevolent fund and to promote a social and sports association commenced. By 1957 the directors felt sufficiently confident of the future of the industry to direct an investigation into the means of establishing retirement benefits, though progress to this end was slow.

Working hours were conditioned by the exigencies of the industry. Wage agreements were based on the national norm of five and a half working days until the statutory five day week was introduced in 1962. One and a half times the hourly rate was paid for overtime and twice that rate for

Sunday working, unless covered by shift allowances. The furnaces, when under heat, were worked around the clock. This necessitated the employment of four shift crews for the melting shop so as to provide for weekend working and the gaps caused by illness or other absences. In the rolling mills there were two shifts, one from 12.15 a.m. to 8.15 a.m. and the second from 8.15 a.m. to 4.15 a.m. A third production shift in the mills was often advocated at times of high demand but was not attempted. Considerable time was required for essential maintenance and repairs to the equipment, without which the levels of output from the existing shifts would have suffered. A third shift would create problems of organisation and costs without commensurate benefit to overall production.

For those in a steel industry, whether occupied in the melting shop or at the mills, operations on the factory floor present an arduous and exacting task. All the more so when, as at Haulbowline during the first era, so much manual operation was required. Physical strength and nerve were demanded. Conditions of constant heat, noise and dust from which there was little escape made work uncomfortable to say the least. There was also danger, familiarity with the risk making it more insidious. During the early years two fatal accidents occurred. There were also a few serious injuries, some alas permanent, when hot metal 'jumped' unexpectedly. These tragic and most regrettable incidents form a sad part of the story of the industry, and yet they had some share in creating the special sense of comradely loyalty, identity and tradition which moulded the workers at Haulbowline into a closely-knit industrial community, with the growing confidence that their achievements and security of employment had given them. Most workers in heavy and intrinsically risky employments such as steel making or deep mining develop this strong community sense, in which pride plays a part, as do their families. After ten years of the struggle for survival in the Irish steel works, the community roots at Cobh and elsewhere in the vicinity were well established; perhaps they were not yet deep enough, but at least it was likely that they would remain undisturbed.

CHAPTER VII

The First Exports

The years 1956 to 1961 were milestones in the history of the undertaking on Haulbowline Island. During that period the company could be said to have grown up industrially and commercially. It emerged from what many viewed as a small local and struggling enterprise into one that became internationally known and respected even though it still remained a tiny operation in comparison with the giants of the world steel industry. It says a great deal for the competence and resilience of the company management, of whom the senior executives could be numbered on the fingers of one hand, that while they tackled the problems of a major capital development on top of the day-to-day servicing of the home market for steel within the Haulbowline range of products, they also undertook substantial exports to many countries.

Exports were not a new idea at Haulbowline. From the outset, particularly at times when home demand was slack, the board had issued directives that export outlets be sought energetically. Occasionally some tonnages were sold abroad. During the Korean War, when world steel supplies fell below demand, many offers for Irish steel were made at attractive prices but the board refused them because it was the company's duty to give priority to home demand. The Irish construction industry was dependent on supplies from the native industry, steel being virtually unobtainable elsewhere at any price. One recalls that an American firm offered to buy the whole Haulbowline annual output at a price far higher than the then home price in Ireland. It is sometimes forgotten how beneficial the existence of an Irish steel industry has been to the economy and to the balance of foreign payments at times of international scarcity.

In 1957 the board decided to engage Countinho Caro of London (an associate of a Hamburg firm of the same name), which dealt in the business of buying and selling on commission steel and steel-making materials on a world-wide basis. They had come to ISH first in 1956 as sellers of black sheet and billets from British and continental works. It is curious to find that the first transactions were marked by complaint and disagreement. From this somewhat unsatisfactory beginning, a relationship of mutual trust and confidence was developed and it was with the help of Countinho Caro and notably their managing director, H. A. Oppenheimer, that exports of steel from Haulbowline became a major activity. The nature of the association as it evolved rested to a large degree on 'gentlemen's agreements' but it was summarised in a letter written by Farren to Oppenheimer in November 1958: 'You are our exclusive selling agents in all parts of the world, except Ireland (including Northern Ireland), and you are authorised to negotiate for our account with any persons or authorities.'

The export policy rested on two assumptions: first, that ISH could produce the tonnage necessary for a worth-while operation and, secondly, that the steel could be produced and sold at economic prices. The first of these assumptions made for a difficult time in the short term because in 1956 the annual ingot output of the existing plant did not exceed 23,000 tons from which the home market had first to be satisfied. When the development project was completed, ingot capacity would be trebled and there would, it was expected, be ample surpluses for export. It was, therefore, important to establish export contacts prior to the development and this was the reason for concentrating so much attention on securing overseas business of a constant character. For this to be achieved, it was essential to establish a reputation for Irish steel as regards quality, fulfilment of customers' specifications and prompt delivery.

As had always been the case since 1947, the capacity of the Haulbowline finishing mills greatly exceeded the ingot capacity of the two small furnaces. To enable export contracts to be accepted with certainty of performance, a quantity of billets was imported for home trade requirements, thus reserving the necessary ingot tonnage for rolling export orders. It was essential to use steel made at Haulbowline for exports in order to satisfy country of origin requirements, particularly where the Irish steel was being sold into Commonwealth countries under the umbrella of preferences granted by the Ottawa Trade Agreements of 1932. After the constitutional change in 1949, when the Republic of Ireland came into being, the Irish steel company when seeking contracts in the Commonwealth had little compunction in arguing through their agents that Irish steel qualified for the same preferences as British steel in these markets. In New Zealand, for instance, much to the astonishment of some United Kingdom producers, Irish Steel Holdings Ltd. of Haulbowline were included in the list of approved suppliers published by the Ministry of Works in Wellington.

The selling prices for ISH exports were necessarily fixed on an ad hoc basis and were not governed by the prices obtainable in the home market. In most cases, the Irish works secured orders for specific contracts offered by public authorities abroad for which Countinho Caro tendered at a price agreed by ISH as the lowest at which business would be acceptable. It was possible to quote keen prices when the order was, in terms of ISH outputs, substantial and involved remunerative extensions of the rolling programmes for the home trade. As a new entrant to the world market, the Irish company was, naturally, prepared to accept small profit margins in order to establish a foothold and win a reputation for its products.

The first major contract was for steel required by the Indian Railways, about 2,000 tons in 1956. It was a difficult specification but was carried through successfully and followed by further orders from the same source. The board minutes record from time to time during the years 1957 to 1960 other export transactions put forward for approval, some accepted and others refused. The reason for the board's concern with these matters of detail was in a sense restrictive. It was not desirable to accept export orders of a kind which could impede or delay the supply of steel for Irish home requirements. Neither was it desirable to do so even for surplus output when the order, although at first view attractive, might be hazardous to execute in terms of technical difficulty or of possible financial loss. These, of course, were considerations of policy applicable to any trading company.

In the years in question, Irish steel was sent to many countries including India, New Zealand, Finland, Jamaica, Singapore, the Philippines, South Africa, Cyprus, Sudan, Nigeria, Kuwait, Denmark and even to Texas. At the same period there was a fairly constant supply of reinforcement to a British firm of bar-benders, running into thousands of tons. Much of that Irish steel is in the bridges and viaducts on the first of the great motorways running north from London.

A spectacular example of these exports concerned New Zealand, the most distant market of all. Towards the end of 1957, the Ministry of Works in Wellington invited tenders for about 5,000 tons of reinforcing steel for a massive hydro-electric scheme on the river Waitaki in South Island, the largest such project yet undertaken in any part of the Antipodes. The barrage was to be an earth dam, the reinforcement being required for the foundations of the generating station, penstocks and bridges. While the specification put forward by the Ministry of Works was unusual as regards the design of the deformations in the steel rounds the ISH technical officers were confident that it could be met and, as the order comprised a big tonnage of bars in two or three economically rollable large sizes, it was decided to make a major effort to secure the business. The government policy in New Zealand was to 'Buy British', and as the Australian steel output was, at that time, absorbed by domestic needs, the Benmore contract was, in effect, at the mercy of the British steel companies, of whom, as it turned out, only a few of the major producers were willing to meet the specification put forward. Thus ISH had a real chance if they were acceptable suppliers and if they could quote competitively.

After consultation with Oppenheimer, who was most anxious to secure the business for ISH, the chairman decided to enlist the support of his friend and colleague in international rugby football, C. S. Hogg, a chartered accountant in Wellington, who, in his role as chairman of the New Zealand Rugby Union, was a well-known and respected figure in his community. Cuth Hogg agreed to act as ISH representative and to pursue the Benmore contract with the Ministry of Works. It may be assumed that his intervention was a factor in providing credibility and perhaps respectability for an Irish steel producer hitherto unknown in New Zealand as a possible source of supply for public contracts. Eventually the Irish tender was accepted, but only after some months of anxiety as to the outcome. In fact it was the lowest tender, but this did not prevent the sales director of one of the British tendering companies from remarking in humourous vein that it was a bit unfair to have to compete with rugby football in New Zealand of all places!

The Benmore contract, later extended to some 7,000 tons, was successfully fulfilled over the stipulated period. Transport was in small Irish coasters from Haulbowline to Liverpool and thence by ocean-going liners to Timaru, a deep water port one hundred miles south of Christchurch. From there the steel was transported by road to the site 6,000 feet up in the Southern Alps. There were no hitches and no rejects, a very pleasing result which was emphasised by further contracts from the Ministry of Works. These concerned two smaller hydro-electric schemes and the viaducts on the Auckland motorway. There were also orders from steel buyers in the private sector. In all about 12,000 tons of Irish steel were shipped to New Zealand during the years 1958–61. (It is equally pleasant to know that the customer is still happy: 'Irish steel is holding Benmore together as we knew it would,' wrote Bob Thomas of the Ministry of Works when sending greetings to the author in July 1978.)

The following extract from the chairman's statement at the twelfth annual general meeting of ISH on 10 December 1959 is relevant:

> While in New Zealand last August, I had the pleasure of seeing great quantities of our steel on the site of the vast Benmore Hydro-electric Project in South Island which is to cost £36 million and produce 478 MW of power. Ministry of Works officials who are executing this project and with whom I had valuable discussions told me that our Benmore contract is the largest single order for steel ever placed by New Zealand. At a modest reception which I gave on behalf of your company in Wellington, we were honoured by the presence of the Prime Minister, Hon. Walter Nash, MP, the Minister of Works, Hon. Hugh Watt, MP, the leader of the Opposition, Hon. Keith Holyoke, MP, and the Mayor of

Wellington, Hon. Frank Kitts, MP (as well as leading steel merchants and contractors). The Prime Minister spoke to the gathering in very kind and complimentary terms about Ireland and Irish steel.

The chairman also reported that during his travels he had visited the site of the harbour works in Singapore at which Irish reinforcing steel was being used 'the quality of which was warmly praised by the technical staff'. He had also visited the Australian steel works at Port Kembla on the coast south of Sydney.

It proved impossible to maintain the New Zealand connection. In 1959 that country was already planning its own steel industry, a small operation based on the use of native scrap, much on the lines of ISH, and this project rapidly came into being as Pacific Steel Ltd. A year or so later the discovery of workable iron sands on the coast of North Island gave promise of a larger integrated undertaking. These developments coupled with the emergence of Japan and Australia as the dominant steel powers in the Far East and the Antipodes, made Irish steel uncompetitive in New Zealand. British steel interests, which were also affected by these events, salvaged something by taking ground floor investments in the consortium established to produce steel in New Zealand. It was not practicable for ISH to seek a similar opening (although the directors considered the point at Hogg's suggestion) because, apart from other policy aspects, all available capital was required to complete the development at Haulbowline, already running well beyond the original estimates of cost.

Apart from the New Zealand situation world steel supplies began to exceed demand quite dramatically and the export of Irish steel abroad became more and more difficult after 1961. There was a revival of home demand coinciding with production problems at Haulbowline, which made it necessary to import bars and sections to fulfil Irish orders. This was a temporary phase but even when that had passed, and the enlarged steel works, after many teething troubles, had been brought into commercial production, such surpluses as existed over home demand could be sold, if at all, only in Britain or in Northern Ireland. On these markets export activity was concentrated.

Quite a useful trade was developed in Northern Ireland, at first in galvanised corrugated sheet and later in reinforcing bars. G. B. Goldsbrough of Hurst and Company, Belfast (another rugby football colleague of the chairman's!) had been appointed ISH representative. Reinforcing steel was supplied for the Belfast wharf extension and for the new graving dock at Harland and Wolff. This occurrence, it was said, moved a reverend member to question in the Stormont Parliament 'why Papish steel from the Republic was being used in a public contract when British steel was available', but this story may well be apocryphal. There was certainly no reluctance on the part of the Northern Ireland steel stockists to buy from Haulbowline when the price was right and many good friends were made in that trade, of whom Bill Simpson of Cohens and Max Hadden of GKN (Northern Ireland) are well remembered. At one time during the mid-sixties ISH were supplying, mainly by sea from Haulbowline to Belfast for warehousing by Cohen and distribution by that firm under merchanting agreements, nearly one-third of the market for galvanised sheets in Northern Ireland. That was about the limit of market penetration to which ISH could go without provoking vigorous reaction from the British galvanisers whose traditional monopoly in the area was being eroded by the intrusion and who had begun to make noises about alleged dumping of Irish steel. After such promising beginnings sales of steel to the North declined for two reasons. One was the growth of demand in the Republic for both galvanised sheets and bars, which reduced export surpluses. The other was the emergence of continental competition at prices which, in some sectors, were below those which ISH were prepared to quote to Northern Ireland stockists or contractors. Good will and satisfactory relations with leading steel purchasers in the North had, however, been well established. Exports to that market continue to be a regular feature of ISH trading even if not constant in volume.

The exports so far described all related to finished steel products whether produced in the rolling mills or in the galvanising plant. Crude steel ingots, the feed stock of the mills, were not normally sold, although several hundred tons had been purchased by an English steel works in Fitzpatrick's time, a catch-penny experiment. In 1964 however, in order to maintain full production and employment in the steel-making shop at a period when the new rolling mills were being installed and tested, export contracts were accepted for large ingots cast to customers' specifications. These differed from those required for domestic rolling at Haulbowline. The trade was substantial while it lasted. The tonnages exported were: to Spain 36,000, to Britain 12,000, to Denmark 9,800, and to Sweden 2,500; 60,300 tons in all. The ingot exports were not the happiest of transactions, troublesome in production and financially unrewarding. This occasioned much worry to ISH management, particularly Farren and E. A. Coleman, who made several visits to Bilbao and Copenhagen in search of solutions to the problems posed by unsatisfactory contracts. Coleman, a certified accountant, who had been recruited from the insurance world in 1956 to the post of secretary, served the company well in many administrative roles, being a pivotal figure in the commercial development of the industry and later in the preparations by ISH for its role as a steel producer within the European Economic Community. A man of considerable talents, business flair and personal charm, he had scarcely yet reached his prime when, tragically, he was killed in June 1972 on a journey to Brussels, in the plane crash near London Airport that brought grief and loss to so many Irish homes and commercial companies.

One further export, this time of a capital nature, deserves mention, if only as a curiosity of the company's history. The small semi-mechanised sheet mill equipment forming part of the assets purchased from the Receiver had, over the years, been the subject of many expert inspections, discussions and plans. It had been installed by ISH on the foundations laid for it by the former company, the missing parts had been provided, and its existence kept alive the idea of rolling sheet for galvanising. There was no answer, however, to the fact that sheet rolling on such equipment had become obsolete and the sheets would prove less uniform in quality and more costly in production as compared with those produced by continuous strip mills of large capacity. It was a surprise when a Turkish steel company offered, in 1965, to buy the sheet mill at Haulbowline. It was sold to them without warranty and for payment of the agreed sum in advance of shipment. The purchase price was many times greater than the historical book value of the original asset but did not fully cover the subsequent expenses in completing and installing the mill. So ended the problem of 'what to do about the sheet mill'. One never learnt what was the operational success or failure of the equipment which Davy Frame had purchased in Belgium in 1937 and which had languished on Haulbowline for nearly thirty years until it found its last resting place not far from the minarets of Istanbul. The disappearance of the sheet mill not only removed a nagging problem, it also left valuable space for other uses.

These ventures in the export field were of considerable importance in the history of the company. It was proved that Irish steel could be manufactured to international standards and requirements and that it could be accepted abroad as of equal quality to steel from any other source. Furthermore, useful experience and contacts in the international steel industry were gained by both commercial and technical officers of the company. Such contacts were to prove valuable later on when the company was called upon to take its place on committees concerned with steel production within the European Economic Community.

CHAPTER VIII

The Second Decade

It has been found convenient, at least for the historian, to separate the story of Irish Steel Holdings during its first twenty years into decades. The first ten years were those of survival and consolidation of the industrial base acquired by the state in 1947. Except for the installation of the pickling, galvanising and corrugating units in 1955, no extension of manufacturing potential or product range had taken place. None the less, improvement and replacement of existing items of equipment to the extent that company resources permitted, had been effected as a continuous process. The second decade, however, was mainly occupied by the planning, execution and disappointments of one major development scheme. There were other proposals for expansion and association to be considered, as well as the measures which the company would need to take for adaptation to conditions of free trade.

The First Development, as it is called in company history, was so important a step in both financial and industrial terms that it seems better to describe it later in some detail. The compressed and indeed somewhat selective chronicle of ten years to which this chapter is devoted will necessarily include occasional mention of the development project but without the particulars and comments which are reserved for the following chapter. Some duplication is entailed but not, it is hoped, to any wearisome extent.

The second decade began in July 1957 and was virtually coincidental with a change in government which brought Sean Lemass once more to the helm of Industry and Commerce. His interest in the progress of the steel works was manifest when the chairman reported to him on 27 July.

Later in the year on 2 October the Minister visited Haulbowline and encouraged the directors to submit rapidly to the government the proposals on which they were already engaged for expansion of the industry.

When the results of the company's financial year to 30 July 1958 came before the annual general meeting in the following November, the chairman was able to report a net profit of £72,000 from home and export trading in conditions 'which were generally unfavourable due to the recession in home and overseas steel markets'. It was emphasised that these conditions still continued. Indeed, production and employment had declined in many steel works abroad. The good news was that the tariff protection for Irish steel had been strengthened by the imposition of minimum duties to offset foreign dumping prices and that Lemass had approved the company's provisional project for development expenditure, as he announced publicly in August 1958.

At this period quite dramatic changes were beginning to affect industry and employment on the shores of Cork Harbour. For ten years the steel works had been virtually alone in this field. Now the construction of an oil refinery at Whitegate on the eastern shore of the lower harbour was begun by a consortium of the main distributors serving the Irish market and, at the same time, the dormant Cork Dockyard Company at Rushbrooke, a short distance up river from Cobh, was to be revived and expanded by Dutch expertise and capital, with active encouragement and assistance from the Irish authorities. Both of these projects used considerable quantities of ISH reinforcing material in the course of construction, but hopes that the adjacent refinery would supply heavy oil for furnace heat came to nought when it was found that the Whitegate products would not include low sulphur oil suitable for steel making. This seeming economic paradox was balanced by another when it became apparent that the steel works could not, in any rational programme of expansion at that time, gear itself to become a supplier of steel plates for the neighbouring industry at Rushbrooke, in competition with the other sources open to Verolme, the Dutch shipbuilders. The huge investment in a modern plate mill of minimum viable capacity, and supported by a large iron and steel plant, could not be justified unless it was linked to the certainty of selling annually at least four times the projected plate requirements of the Verolme Cork Dockyard and of doing so profitably against the competition of existing foreign plate mills hungry for business. All economic sense and logic was against attempting such a large venture or even recommending it to the government for serious consideration.

The emergence of the Dockyard project did, however, raise another problem which greatly agitated the directors of ISH during 1958. As part of their plans, Verolme proposed to use the graving dock at Haulbowline for ship repairing facilities ancillary to the main operations at Rushbrooke. Both the steel company and the naval services offered sour comments on this part of the scheme, claiming that a third major user of the harbour and dockside facilities would involve unacceptable congestion and interference with existing activities. ISH were particularly concerned over the effect such circumstances would have on their development plans and commercial prospects. However, the IDA and Industry and Commerce were anxious to place no obstacle in the path of Verolme and moreover, as ISH found when they met him, Cornelius Verolme was a very dynamic personality whose confidence and wishes were difficult to oppose, so that it seemed for a while that ISH would have to bow to the inevitable intrusion. As it turned out wiser counsels eventually prevailed and Verolme did not exercise their option to use the graving dock and were content to expand at Rushbrooke where the dockyard operations have been maintained successfully. Over the years, in matters of mutual concern hopes expressed in 1958 for good-neighbour relations between the two companies have been fulfilled and in recent times have been marked by the fact that the chief executive of each company has been appointed to the board of directors of the other.

In considering the financial arrangements for a major development in steel production at Haulbowline the directors gave thought to the possibility of external participation which would relieve the state of some part of the heavy capital expenditure involved. The chairman discussed this aspect with the Minister in the context of the latter's encouragement to proceed with development. It was a theme which was to recur on other occasions during the next twenty years. In 1957 the view of the board was that Irish control of the

Foundations being prepared for new melting shop, 1961

Irish-made steel in hydro electric scheme at Benmore, New Zealand, 1960

11. Visit of Taoiseach, Mr Sean Lemass, in 1963, to mark completion of first major development plan involving the installation of Maerz open-hearth furnace, section mill and bar mill

12. Visit of Taoiseach, Mr Sean Lemass, in 1966, to open the new Haulbowline bridge. Two future Taoisigh, Mr Jack Lynch and Mr Liam Cosgrave, are

13. Visit of Taoiseach, Mr Jack Lynch, to Haulbowline in 1973. Also in the picture are Mr D. Frame, Mr G. P. S. Hogan and Mr S. O'Leary, Lord Mayor of Cork

14. Visit of Minister for Industry and Commerce, Mr Justin Keating, to Haulbowline, 1975

15. Pouring of molten steel from electric arc furnace into suspended ladle

16. Artist's impression of the future appearance of Haulbowline when the modernisation and development scheme is completed in 1980

steel industry should be maintained but that a suitable partner could be admitted to assist in development. There should be no question of 'hawking' the industry for sale to foreign interests. The tests of suitability for partnership should in general be:

(*a*) the partner should be content with a minority interest:

(*b*) the partner should be equipped to provide advanced steel technology and research facilities in support of ISH management;

(*c*) the partnership arrangements should be such as to promise marketing of increased outputs;

(*d*) the partner should be prepared to provide a substantial amount of risk capital as assurance of his continued interest in the Irish industry.

Lemass agreed generally with these propositions. Asked whether it was likely that such a partner could be found, the chairman said he would try, but it seemed probable that such an association would be more attractive to a potential partner after the first development had been completed and was operating profitably. Moreover, ISH would then be in a much stronger position to make a satisfactory agreement. With the Minister's encouragement, the chairman sounded Sir Ernest Lever of Richard Thomas and Baldwins. He also discussed the question with Harald Peake, chairman of the Steel Company of Wales (SCOW). In both cases the results were negative, due not to lack of interest, but to the impracticability of considering an Irish investment when their companies were so heavily engaged in modernisation and expansion of their own plants. A similar response came to tentative enquiries in the USA and Belgium. The visit to Lever was not altogether unfruitful. He agreed to loan one of RTB's experienced managers to ISH for technical advice and refresher training of the galvanising operatives at Haulbowline.

The question of reinforcing technical management at ISH had become important on the threshold of major development. An unsuccessful effort was made to recruit an experienced development engineer from the steel industry abroad. Farren also made contact with the British Iron and Steel Research Association (which ISH joined as an associate member) so as to have access to their technical and research information. The senior technical staff at Haulbowline was effectively

Radley and Warbrook, both able and dedicated men, but heavily pre-occupied not only with the formulation of development plans but also with current production problems in support of the export drive. In the latter part of 1957, Everett Timmerman, an elderly retired officer of the Steel Company of Canada, came to live in Ireland and was introduced to ISH from government circles. He was eventually employed until February 1961 as a part-time consultant, and he took part in the work of the committee chaired by O'Shea which produced the first development project for submission to the Minister. Timmerman was enthusiastic in offering ideas and comments but his contribution was limited by ill-health and unfamiliarity with the peculiarities of Irish industrial life. It was, however, his presence and his Canadian connections that brought about the curious episode connected with the name Premium Iron Ores, which during 1959 and 1960 diverted some of ISH attention from its own development programme.

Early in 1959, Patrick Cavanagh, a vice-president (later president) of Premium Iron Ores, Ltd., of Montreal came to Ireland with a proposal that his firm should establish a plant at Haulbowline which, using Irish ores or residuals as the basic ferrous material, would supply cold or molten iron to the ISH steel-making furnaces. Premium were a well-established company, with associates in the Canadian mining industry. According to Cavanagh's own statement 'their business was in the sale of ore rather than in its processing' but their competence to install an ore reduction plant was never in question. Naturally the prospect of using Irish minerals to support the steel industry held great attractions and Premium were encouraged by Lemass and by the IDA to press ahead with their plans. The ISH stance was that if Premium could supply iron of the right specification at an economic price, their development project could be adjusted to this situation, but that they could not accept responsibilities in relation to the Premium plan other than as a purchaser of the end product.

Dr Auger, the chief geologist of Ungava Mines, carried out surveys throughout the Republic on behalf of Premium which showed that only at Ballineen and Ballard in the Avoca area were significant quantities of iron ore present and even there economic extraction was a dubious proposition (a finding relevant to the question posed in the first chapter of this book). Success depended, then, on the processing of the ferric residuals from the Avoca pyrites which were being mined for use in fertiliser manufacture at Arklow. The discussions drifted on, Premium probably feeling that ISH were holding back, while ISH maintained, as did the IDA, that precise and detailed Premium proposals on which final judgments could be formed did not exist on paper. Eventually, to advance matters, at a conference in September 1960, ISH offered to buy annually 50,000 tons of iron briquettes or molten iron from the proposed Premium plant for approximately £19 per ton. Writing to the Department of Industry and Commerce at the time, the ISH chairman said that in view of the many unknown factors caution would have suggested a lower price but his board 'had decided to offer £19 per ton as a very significant contribution towards enabling the project to proceed to a satisfactory conclusion. We are not offering that price on a basis of horse-trading but as the limit to which we could safely go.' Cavanagh replied on 15 November that, having studied the matter, Premium was not able to consider further supplying ISH with molten iron at the price offered and that they were examining the possibility of processing Canadian iron ore in Ireland. In spite of reminders from the IDA, nothing further was heard from Premium and an exciting prospect died. One wonders if it had ever been really alive. ISH had their doubts and were glad to see the end of so tenuous a proposition which might have imposed onerous conditions on the steel industry. There was relief also in naval circles because the proposed iron reduction plant on Haulbowline would have deprived the navy of essential space and facilities. Indeed, it was suggested that the Naval Service would have to move elsewhere at an estimated cost of £1 million.

The case of Premium Iron Ores has been described in some detail because it was of a pattern with other incidents that occurred in the years that followed. Encouraged by the IDA and well aware of the grants and tax reliefs open to exporting industries established in Ireland, foreign companies or their agents came to investigate possibilities, including the steel industry. Even the least promising ideas had to be examined and discussed with the visitors, often leading nowhere but involving much time and trouble, sometimes over a protracted period, for the managerial and technical staff at ISH, usually overburdened by day-to-day problems of production, marketing and finance. Not all of these interventions were fruitless; some proved stimulating and established new ideas and connections, even where worthwhile commercial results did not ensue.

In the early years of the second decade the steel industry continued to make progress. At the annual general meeting of the company in December 1959 it was possible to report record production and a net profit of over £80,000, even though selling prices had been maintained unchanged during the year. Exports had been doubled, 'this being no doubt aided', the chairman said, 'to some extent by the recent strike in the United States steel industry but there are encouraging signs of a gradual and we hope permanent increase in world demand for steel'. At this meeting, the chairman's statement included, in the first of many such exhortations, a plea for understanding that a continued high level of employment depended on the industry's ability to control costs, and hence that moderation should be exercised in relation to wage claims.

A year later, the chairman again announced record production figures even though operations had suffered some interference from the progress of the development works. Turnover was over £2 million yielding a net profit of £87,500. Exports had declined due to pressure of increased home demand. The accounts showed an issue of shares (£300,000) to the Minister for Finance, this being the first instalment of the capital required for the development project. Thus, for the first time since its establishment in 1947, the company received a payment from the Exchequer, a circumstance not often realised by the general public who tended to regard all state enterprises as constant recipients of aid and a burden on the tax-payer.

The board minutes of this period (1958–61) indicate constant preoccupation with a few major topics: the progress and problems of the development project; the decisions regarding the quantity and destination of export sales; adjustment of selling prices to market conditions; questions of recruitment and remuneration. Some items in these minutes may show a growing sense of the need for promoting greater public awareness of the industry. In the early days, the directors had an almost paranoic aversion to publicity believing that no news about the industry was better than untimely or ill-informed comment and premature proclamations of achievements to come. A sales manager, whose duties naturally included sales promotion, was appointed. A firm of advertising and public relations agents were engaged. The company participated in Cork industrial parades and exhibitions and later at the Dublin Spring Show. The brand name *Steelile* for the company's galvanised sheet was registered by the Patents and Trade Marks Office, as also was the company's general trade mark or symbol, the 'little man' holding aloft like an Olympic champion a heavily weighted steel bar, together with the slogan *ní neart go cruach* testifying to the great strength and durability of the product. The latter device was designed by Tebb, the foreman fitter, who received a modest gratuity of £30 when the company decided that his suggestion was meritorious. That was in April 1961 and as Tebb retired shortly afterwards and went abroad, possibly he never saw his imaginative offering being put to use.

It was in this period too that the company's thoughts were directed towards the need for a road bridge between Haulbowline and the southern mainland. As early as March 1958 the board had asked William O'Connell of the Cork engineering consultancy firm O'Connell and Harley, to give them a 'guesstimate' of the cost of such a bridge. It was obvious that if any such proposition were to be feasible, the company must first, before any ideas for a bridge became public, acquire property on the mainland. In February 1960 the area known as Paddy's Block, a peninsula jutting out from the southern shore towards Haulbowline and comprising thirteen acres of dunes, was acquired for £700. A year later some 45 acres of farmland behind the peninsula came on offer. While this area was not essential for the bridgehead, its acquisition would, in the long run, offer great possibilities for ISH either as a site for its own extended activities or for other industries dependent on steel for their manufacture. Using intermediaries so as to conceal the interest of ISH which could have rocketed the price, the chairman, with his board's authority, concluded the purchase of the area for about £12,000. These transactions were not only an essential preliminary to tackling the bridge project, they also added significantly to the assets of the steel company. Three years afterwards, in 1966, the Department of Defence willingly transferred to ISH for a token sum of £150 its proprietary rights in Rocky Island. This islet was a vital link in the design of the long road bridge.

In 1961 there were changes in the structure and personnel of the board of ISH. In March Hogan, with the approval of the Ministers for Industry and Commerce, and Finance, retired from his civil service post and entered into an agreement with ISH to fulfil the duties of an executive director in addition to those of his position as chairman of the board. In September, Jack Lynch, who had replaced Sean Lemass as Minister for Industry and Commerce when the latter succeeded President de Valera as head of government, spoke to the chairman of his intention to appoint additional directors. This was a courtesy which Ministers usually extended to chairmen of state companies, before an appointment was made. The Minister's selection clearly added strength to the board and was also welcome because local representation in the direction of the industry seemed overdue. The new directors were from Cork. Fred J. Cross was long preminent in the motor trade in that city and had engineering qualifications, while Dr Tadhg Ó Tuama was an industrial scientist of high repute, for many years engaged in the sugar industry and the leader in that company's research and development programmes. Both of them remained for too short a time with the steel industry, but while they were there they contributed, from their abilities and varied experience, much encouragement and good sense throughout one of the company's most difficult periods. These appointments brought the board to the full number permitted by the then Articles of Association.

In December 1961, the annual accounts showed a fall in net profit to £56,000, although turnover had been maintained at around £2 million. In January 1963 it was reported that net profit had again declined to £43,000, while due to buoyant home demand only £200,000 of exports had proved possible. At that annual general meeting, emphasising the magnitude of the transition problems facing the company, the chairman said, 'This development is virtually the creation of a new industry and not merely a job of extension or enlargement.' A year later the financial position had significantly worsened. Production in the rolling mills had ceased for a protracted period, longer than had been foreseen at the planning stage of development. Notwithstanding increased outputs and sales of galvanising sheet, the company's revenue fell by 15 per cent while cost in all sectors increased. A net loss of £257,000 was shown in the accounts, and was covered by a transfer from the company's general reserve. The chairman was pessimistic about the following year, complained of competition from 'dumping' prices and urged that in this period of transition ISH should continue to receive adequate protection. Citing examples from abroad he said, 'I know of no steel producing country which does not now protect its steel industry, and many of them, including the Common Market countries, are currently seeking increased protection.' Emphasis was added to that comment when not long afterwards a surcharge of 15 per cent was imposed on steel imports into Britain, which damaged ISH exports to that country.

It was no surprise that another substantial net loss – £191,000 – appeared on the accounts presented in December 1964, again covered by a transfer from the general reserve, now virtually exhausted. It was stated that the loss was 'almost entirely occasioned by exceptional expenses arising in the course of the installation and initial operation of the new mills'. Such expenses had also been a factor in the previous year's net loss. At the directors' request, the government agreed that a sum to meet these dead-weight expenses should be included in the state capital subscription towards the cost of development.

At the end of 1964 the company's balance sheet showed £5,350,000 for shares subscribed by the Minister for Finance within the maximum of £6 million permitted by legislative authority. A further long-term loan (twenty-five years) of £150,000 had been obtained from the Irish Life Assurance Company, thus doubling the obligations to that institution but *pro tanto* reducing overdraft indebtedness to the ISH bankers. In the same balance-sheet, company resources tied up in stocks reached the high figure of nearly £1.25 million due to accumulation of scrap steel while the new furnace capacity was under construction and also to the effect of a protracted strike in the building trade which caused ISH, willy-nilly, to hold 2,500 tons of finished steel at Haulbowline until the customers were able to accept delivery.

In spite of all its difficulties, including persistent teething troubles in the new rolling mills and poor productivity ratios, the industry made progress in 1964 and 1965. Good performance in meeting the growing demand for galvanised sheet offset to some degree the losses in other sectors. The new Maerz furnace also gave satisfaction. In the financial year ended June 1964, turnover was raised to £3 million including exports sales of £400,000. In 1965 these figures were again increased to £4.35 million and £1 million respectively and a small overall net profit was secured which would have been greater but for the effect of two external industrial disputes (oxygen and transport) which curtailed production and added to costs.

The outlook was, however, still sombre as the chairman reported to the annual general meeting in December 1965. All material costs and wages were rising and market prospects suffered from a general recession in the Irish economy and consequent decline in orders from stockists and manufacturers. At the same time, the international steel trade was weak and growing more so. Surplus steel flooded world markets at artificial prices and even the great USA economy protected its steel industry with high tariffs. Thus ISH faced greater difficulties in maintaining exports, and such orders as could be secured were at exigent prices. In the previous year ISH had paid £50,000 in British import levies to maintain its sales, mainly of ingots, to that market. The real crux, however, continued to be the growing awareness that the plans on which the First Development was based eight years earlier had been overtaken by events and that new measures and strategies would be necessary if the policies and intentions in the government's Second Programme of Economic Expansion were to be fulfilled by the steel industry. As often before, the chairman had called for hard work and restraint in seeking its rewards. 'The actions of men and not the performance of machines,' he said, 'will determine the success of this industry.'

Thus it was that in December 1966 a further net loss (about

£66,000) was reported. Turnover fell by 11 per cent but due to remarkable efforts by the general manager and his staff, with the zealous aid of the export agents, exports were maintained at the high level of £950,000. There was a welcome announcement of financial recovery a year later when sales revenue once more exceeded £4 million, even though exports declined to £700,000. A net profit of £182,000, the highest in the company's history, was the result of record outputs from the plant and improved productivity. The balance-sheet at 30 June 1967 showed issued share capital of £5.785 million (the cost of development) but the company's liquidity was healthy, current assets exceeding current liabilities by nearly £1.4 million.

In financial terms the company's situation had altered quite dramatically during the decade. A comparison of the report and accounts of 1957 with those of 1967 illustrates this. In the former year the total of the balance-sheet was £561,000; ten years later it was £6.7 million including fixed assets standing at nearly £4.7 million (after depreciation). In 1957 the fixed assets (net) had been only £187,000. By the end of the decade sales revenue had quadrupled as had salary and wages costs even though the number employed had increased by no more than twenty-five per cent, a sure sign of the inflationary trends in the Irish economy, which became a growing problem towards the end of the period.

There are some other features of the decade which deserve mention. One of these was the emergence of regular meetings between company officials, usually headed by the executive director, and members of the committee of the National Association of Iron and Steel Merchants which represented the interests of the company's main customers in Ireland, the merchant stockists. Discussions with that body had, it is true, taken place in the early years of the company but these occasions occurred usually in times of crisis and sometimes in a less than harmonious mood. It was essential, as the market developed, for the company to have information from the trade as a whole regarding its probable requirements for a year ahead so that rolling programmes and company purchases of materials could be sensibly and economically planned. Equally the trade sought notice of any contemplated changes in company plans and policies. Considerable progress was made in establishing good relations and procedures and in avoiding confrontations. The understanding and cooperation generated by these conferences were often gratefully mentioned in the director's statements to company general meetings. On the trade side much of the credit for the progress made should go to the efforts of a succession of able chairmen of the National Association, notably George Gamble (Brooks Thomas), Ted Usher (Heitons) and Franz Rothschild (Ferrum). Sometimes, indeed, the meetings were also a forum for complaints or criticisms by customers and sometimes the company had a bone to pick. Whatever the difficulties and however deep the cut and thrust of sharp business exchanges, good relations always triumphed, while occasionally goodwill and personal friendships were cemented at pleasant social evenings in Cork following the long meetings.

The National Association was not the only representative body in the steel trade to which the company devoted attention. Valuable interchanges and cooperation were also secured from meetings with the Structural Steel Fabricators Association and, at a later period, with the separate body set up to represent Provincial Steel Stockists. During the time of preparation for the entry into the Common Market the company took its place on the Adaptation Council for the Construction Industry but, it must be admitted, without any significant advantage to the interests of ISH.

In June 1962, on the intiative of Dr Ó Tuama, who had taken an interest in industrial relations at ISH, the board considered the possibilities and advantages of a standing committee in the works composed of representatives of both management and labour. The sugar factories were pioneers in this field in Ireland and their model, of which Tadhg Ó Tuama had had favourable experience, was adopted by ISH subject to modifications required by conditions in the steel industry. Even though the scheme excluded questions of wages and other conditions of employment from the agenda of such a joint committee, the idea was coldly received by the unions at Haulbowline. The ITGWU proved the most difficult to persuade and it was not until November 1964 that they agreed that their members would participate in a Joint Management/Labour Committee under its constitution as approved by the ISH Board. This delay of two and a half years between concept and fulfilment, solely on account of union misgivings, was frustrating and, in the light of subsequent experience of the working of the Committee, seems somewhat inexplicable. The Committee is a partly elected (the majority) and a partly nominated body under the chairmanship of the general manager. Basically it is intended as a forum for two-way communication between management and workers as to the problems and performance of the industry and the means of improvement. Its main success has been the creation of a common identity and purpose among all who work in ISH and the flow of ideas from the industrial floor. A general meeting of all employees is held each year to which reports from the Joint Committee are submitted.

Early in 1962 ISH entered into discussions with the National Fence Manufacturing Company of Bladensburg, Maryland, USA. This was a private company successfully operated by two brothers named Long, of Irish extraction, who decided to expand their business into Europe. When Clyde Long came to Haulbowline he was introduced by Father William Hogan, S.J., head of the department of Industrial Economics in Fordham University, who had a world-wide reputation as economic analyst and forecaster in the field of steel production, and whose advice was often sought by the leading steel corporations in the USA. (He is the author of the standard History of the United States Steel Industry, an immense work running into many volumes.)

The project was that National Fence should erect a factory beside ISH using Irish wire rod and producing chain link fencing, farm wire and concrete reinforcing mesh. Naturally the directors welcomed the proposal, which was also encouraged by the IDA, and they undertook to give a long lease of a site of up to twenty acres on reclaimed land on the eastern side of the island. While rejecting the suggestion that they should invest substantial capital in the cost of the plant, ISH erected at their own expense a building to house the pilot plant which Clyde Long said he needed to make progress. After a promising opening negotiations began to limp. National Fence may have met unforeseen difficulties in arranging to implement their scheme. After a last conference at Haulbowline in June 1964, the affair petered out.

This was not the only occasion when Father Hogan took an interest in the Irish steel industry. Afterwards he came to Haulbowline on several occasions, when his comments and advice were practical and helpful. One of his suggestions was for the further modernisation and enlargement of the bar rolling mill and indeed, he sponsored a proposal for association in such a scheme between ISH and one of the lesser American steel producers. Although tentatively discussed, nothing came of this as it coincided with a different plan for external participation referred to later in this book as 'Project Ivor'.

During a visit to Canada, MacCarthy, secretary of the Department of Industry and Commerce, had made contact with some people in the steel industry there. Following discussions in December 1962 between the chairman of ISH, MacCarthy, and Whitaker of Finance, it was agreed that two steel managers from the Steel Company of Canada would visit Haulbowline and furnish a report of their findings. Thus Heffernan and McPherson came in March 1963 and duly reported. They were engineers and mainly concerned with technical problems. What they had to say confirmed ISH's own views regarding the execution of the development project and the reasons for the failure to achieve production targets.

A further intervention came in August 1963 in the person of the remarkable Jack Mulcahy, who twice visited the chairman when the latter was on holiday in Kerry and expressed his willingness to help, particularly in the procurement of expert advice from the US. Jack Mulcahy and his own firm, Quigley Inc. of New York, held world-wide patents for a new process of manufacturing and applying roof-chrome (a substance which prolongs the life of furnace brick linings and hence confers economies in steel production). He had undertaken the establishment of a subsidiary factory at Cork which would export the Quigley product to Britain and Europe, this being only the first in a number of enterprises which he launched in the land of his birth. Mulcahy was aware from conversations with the Minister for Industry and Commerce that ISH had problems and it was this circumstance that made him seek out Hogan with an offer of such assistance as he could furnish from his wide contacts in the American steel industry. Discussions led to the arrangement that Frank Gregory of Republic Steel and a colleague would visit Haulbowline in the near future and advise on steel-making practices. The chairman, in accepting Jack Mulcahy's warmhearted and altruistic intervention, tried unavailingly to establish that ISH would meet the expenses of any technical assistance on a commercial basis. That was not the Mulcahy way; it was useless to argue the point on that occasion or when subsequent visits of steel experts enlisted by him took place. In parenthesis it can be related that these visits to Waterville were the first sight Jack Mulcahy had of that lovely place. Its future was profoundly changed when he invested large sums in establishing at that resort a luxury hotel and a championship golf course. It may be reckoned a failure by the chairman of ISH that he did not attract these millions to the steel industry!

Frank Gregory's visit was certainly beneficial as he made useful suggestions to the ISH general manager and his melting shop staff. Subsequently Mulcahy suggested that a man whom he had chosen for a post in his selling organisation in Europe should instead go as works manager to Haulbowline. 'Your need is greater than mine' was the basis of the offer. While acknowledging the kind thought, the chairman explained that in the case of a state company, any appointee must pass through the ordinary process of selection after application for employment and interview. And that was done.

It was in these circumstances that Ian McGurk, currently employed by the Steel Company of Wales (SCOW) as a manager in one of its departments, was appointed works manager at Haulbowline in January 1964. For some time the general manager had been urging the board to strengthen senior supervision in the works where too few people were sharing too many tasks and where some new technical blood was needed. Warbrook's transfer to whole-time work on research and development left a vacancy for works manager. For the needs of that post McGurk had considerable potential and melting shop experience. Support in other sectors of the industry was required and John Borrie, a youthful rolling mills manager with Workington Steel Company, was recruited for the post of assistant works manager. A new chief engineer, from the British steel industry, was appointed in 1965. This was Geoffrey Bullivant who was aided by two Irish assistant chief engineers, Reynolds (electrical) and

Kennedy (mechanical). Ian McGurk's service as works manager came to an end in June 1965, however, when he offered his resignation unless he were given a greater degree of autonomy. What he sought could not be conceded. Borrie, rather prematurely, became works manager and his assistant's position, after wide advertisement, was filled by a Scot, McKenzie, who took charge of the melting shop. Bullivant (1968) and Borrie (1971) eventually resigned to take up appointments abroad. Following Borrie's resignation McKenzie also left. The choice of works manager, as a temporary expedient, fell on K. C. Sharp, who had many years of managerial experience with SCOW. He retired in January 1973 making way for John Rafferty who had been acting as deputy works manager. Thus a period of nearly nine years of hit or miss reliance on expertise and experience from outside Ireland in filling senior posts in the works came to an end. The new team of home talent, two of them old and tried hands, took up the reins under Farren's overall authority as general manager, Rafferty (works manager), Radley (rolling mills manager), Warbrook (planning controller), Reynolds (chief engineer). It is not recorded whether anyone recalled the recommendation of the original directors twenty-five years earlier, albeit in different circumstances, that it would be better for the industry to grow its own staff than to rely on recruitment from abroad.

The end of the decade also saw changes in the composition of the directors. In December 1965, Tadhg Ó Tuama, much pressed by his other engagements and under strain, was obliged to leave the steel board. In August 1966 John Connor resigned following his appointment as chairman of Aer Rianta, also a state company. At the end of that year Tom O'Shea claimed the privilege of age and retired. In the following July, Fred Cross pleaded ill-health as reason for his resignation, alas only too justifiably, as the good man died before a year had passed. These events represented an effusion of blood from the company which these men had all served with ability and dedication. In O'Shea's case his departure meant the end of an era. No man who had given so much dedication for so long and was so much loved could ever be replaced. In more than one sense he was the life and soul of Irish Steel.

Bernard Devlin, a principal officer in Industry and Commerce where he had been dealing with ISH affairs, was appointed to fill one of the vacancies in December 1966. In June 1967, D. D. Frame, chairman of the Hammond Lane group of companies, was invited by the Minister, George Colley, to join the ISH board. David Frame had been a director of the original Irish steel company founded by his father. His desire to promote the interests of the steel industry was undoubted as shown by the friendly help and advice he had given to ISH many times when they were faced with scrap supply problems. Thus when Fred Cross retired, the board was again reduced to three, Hogan, Frame and Devlin.

As early as 1956 the board minutes record interest in the establishment of a modest Dublin office. This became essential in 1961 after the appointment of the executive director, in order to form his headquarters when not attending at the works. Rooms were found, originally at 41 Fitzwilliam Square and, from 1964 onwards, at 6 Mount Street Crescent. Meetings of the board, which are sometimes held in Dublin, and other meetings and conferences could take place at the office which also provided a convenient base for company officials visiting Dublin on business. Assistance in the Dublin office was ably provided by Miss Anne Devlin (later Mrs Binley) who served ISH well in that rather isolated post for thirteen years. In 1978 the Dublin base was moved to 25 St Stephen's Green, where the chairman and also Gerry Farren have their offices and secretarial staffs.

During the closing years of the second decade, a Scottish company established by two former departmental managers of Colvilles, the Clydeside steel works, was employed to extract the metallic residuals from the slag poured off at each tapping from the open-hearth furnaces, and to return the metal so recovered for use in future charging of the furnaces. The contract obliged Slag Systems Ltd. to provide the necessary labour and equipment, both for transport of the slag from the melting shop to the dumping ground and for the separation process. These contractual arrangements achieved marginal economies for the steel industry. The principals of Slag Systems, Aitken and McLay, the former with expertise in steel making and the latter in steel rolling, offered consultancy services at the period when the board had directed that the adaptation of the industry to meet conditions of free trade should proceed urgently. McLay and Aitken were engaged on a retainer basis and a joint committee, composed of them and the ISH technical staff, produced a scheme for the early expenditure of £5/600,000 on improvement of existing plant and facilities, and for a longer-term project intended to raise the works output to 250,000 tons a year. In December 1965 the directors submitted these proposals to the Minister for Industry and Commerce together with their own comments on the state of the industry (admitted to be 'unsatisfactory'), on the future prospects and on the options open to the government. A meeting which the chairman and Farren, together with their technical advisers including Aitken, had with the departmental officials in March 1966 proved inconclusive, except with regard to agreement on the need for some of the early remedial measures proposed by the company. The major proposals, however, fell into abeyance following the emergence of a project for association between ISH and a leading company in the British steel industry.

This did not take place immediately but was the sequel to a visit in June 1966 by the chairman of ISH and the general manager to the steel works at Brymbo near Wrexham in North Wales. These works, owned by Guest Keen and Nettlefold, were of moderate size and produced from open-hearth furnaces steel for rolling into forging-quality billets, a product which ISH had in mind as part of their scheme for adaptation. To inspect these operations was the purpose of the visit, an invitation having been arranged through the good offices of Paul Conlon (appointed a director of Irish Steel in December 1978) of the Irish fertiliser industry, a former executive of GKN, Cardiff, who accompanied the party to Brymbo. In private conversation with E. Davies, the general manager of the works, the ISH chairman talked about his company's attitude towards external technical assistance or even external financial participation in planning for the future. The hints did not pass unnoticed. In the following September, Sir Douglas Bruce-Gardner, a senior member of the GKN main board, accompanied Emrys Davies on an informal return visit to Haulbowline. In November 1966 exploratory discussions took place at Haulbowline between

Hogan, Fred Cross and Farren for ISH and Bruce-Gardner and his co-director Lionel Pugh for GKN when, without commitment on either side, it was agreed that a plan for expanding the Haulbowline works would be worth further examination after detailed technical and commercial studies had been carried out on a specific project. Following this meeting, Hogan, with the approval of the board, wrote to Bruce-Gardner requesting GKN to carry out a feasibility study for which ISH would pay a fee. What transpired subsequently falls into a later phase of the story and is recounted in chapter ten.

In a lighter vein one may raise a smile by recalling some references in the board minutes of the early days. There are indeed few such piquant entries in the dry chronicle of great and small events. On two occasions the minutes refer to the management horror at the 'obscene language' used by a Western customer and to the decision to refuse him further supplies until he apologised. It is not recorded whether he ever did so. Another form of sensitivity is shown by indignant consideration of alleged defamatory statements regarding the industry, once by a retired legal luminary in a letter to a newspaper and another time by a Chamber of Commerce in its gazette. Letters by the ISH solicitor produced, in one case, an ironic withdrawal and in the other a published apology and the solicitor's costs. Many years afterwards when there was wrath about another libel, an eminent Senior Counsel wisely advised the chairman that no state company would be likely to gain a jury's support in proceedings for defamation unless the outcome were to be the insult of one penny damages. One other item in these old minutes suggests the directors' hunger for some little scrap of praise when they thought it worth while to record in August 1957 that a letter had been received from a civil engineer, in Kuwait of all places, congratulating the company on the quality of its steel. *O si sic omnes!*

A momentous event in the second decade of the industry was the official opening of Haulbowline Bridge on 24 October 1966. The opening ceremony was performed by the Taoiseach, Sean Lemass, and the bridge was given an episcopal blessing by His Lordship, Dr Cornelius Lucey, before a large attendance and in the sunlight of a fine autumn day, which displayed to perfection the attractions of Cork Harbour and the graceful proportions of the long bridge designed by the late Cyril Roche, B.E. The day was made memorable by the presence of the Taoiseach, to whom, as a Minister, the steel industry owed its inception and subsequent growth, and who, although this was as yet unrevealed, was then rendering to it his last act of support and inspiration. His gesture was not completed until, at a public luncheon in Cork, to which the company's distinguished guests came to celebrate the occasion, he spoke about the steel industry, its achievements and his hopes for the future. When, in a day or two, it was announced that he was resigning from his position as head of the government and retiring from active political life, the steel community at Haulbowline remembered his support with gratitude and took some pride in the fact that he had come to open the bridge when preoccupied with such grave decisions. By the direction of his board, the chairman wrote to Sean Lemass conveying the sentiments of all in the company and received the following reply:

> Dear Mr Hogan,
> I am deeply appreciative of your kind message on the occasion of my retirement.
> The fact that my last public appearance as Taoiseach was my visit to Cork for the opening of the Haulbowline Bridge is a matter of great satisfaction to me. I hope that Irish Steel Holdings will continue to make good progress.
> Mrs Lemass joins me in sending you our good wishes for your own health and welfare.
> Yours sincerely,
> Sean F. Lemass

CHAPTER IX

The First Development

The plans for development of the industry on which the management at Haulbowline were engaged during the second half of 1957 and the early months of 1958 can be observed against a background of dramatic expansion of steel making throughout the industrialised countries. World production of steel had increased from 190 million tons in 1950 to 300 million tons in 1957. This upsurge had been achieved not only by enlargement and modernisation of existing plants but also by the installation of a number of new giant steel works designed to meet the projected rapid growth in demand for steel products of all kinds. As compared with the situation before the Second World War, consumption of steel per head of population by 1957 had grown by 100 per cent in the United States and 60 per cent in Germany and the United Kingdom.

In Ireland circumstances were different because of the slow development of the industrial sector of the economy. Consumption of steel had increased from a low base by not more than 15 per cent during the early post-war years but it was reasonable to assume that the annual rate of growth in demand would rise as economic activity developed in response to government efforts to promote both agriculture and industry.

It was almost an article of faith in the world steel industry at that time that the future lay only with very large fully integrated works where the immense capital investment would be rewarded by economies of scale, and that there would be little, if any, place for smaller works except when engaged in producing specialised products. The exceptionally small industry at Haulbowline, not integrated and not producing special steels had, accordingly, a somewhat bleak future, as many critics of the proposal to incur expenditure on expansion were not slow to assert. The directors of ISH and their management were prepared to take a more optimistic view of the prospects for the industry, provided that development was regarded as a first step, involving a moderate amount of capital and with assurances of tariff protection on the home market while the expanded industry found it feet. If this first step was successful, further development could be considered with the object of eventually establishing an integrated steel works where pig or molten iron would be produced from imported ore. It was admitted that the projected development with its limited range of product would not have the capacity to supply much more than a third of the country's total steel requirements. That range was, however, important for essential needs of agriculture and of the construction industry. Moreover, a strong element in the case for preserving the nucleus at Haulbowline was that these needs could continue to be met in times of international scarcity or of war situations such as had happened during the Korean crisis of 1950–52. Preservation of ISH meant expansion; the industry could not stand still.

On these premises the plan which emerged from the protracted studies by Farren and his technical staff was, after endorsement in a report signed by the directors, submitted in March 1958 to the Minister with a caution that the estimates were necessarily tentative. The total expenditure was taken as £2 million. Projected output of finished products for sale was 53,000 tons a year. In August 1958 the government approved these plans in principle but stipulated that no commitments involving state investment should be made until final proposals had been submitted and confirmed.

The directors decided that acceptable final plans could not be prepared without assistance from consultants experienced in such work and in the vetting of tenders received from prospective suppliers of plant. After much consideration the choice fell on John Miles and Partners, London, who had experience of installations and extensions at smaller steel works and were recommended following enquiries made through the Irish Embassy in London. (Although it was not known until recently, Miles had been recommended to the

Minister for Industry and Commerce in 1946 as suitable consultants in relation to a suggested plan to cure the ills of Irish Steel Ltd.)

Following discussions, the consultants were commissioned to prepare a general technical and economic report with recommendations as to the best means of achieving the desired expansion by ISH. Their report was received in March 1959. It was in general an endorsement of the company's proposals but differed from them in some material aspects, notably as regards the larger size of ingot to be cast, the need for a medium section mill independent of the blooming mill and the addition of wire rod to the production range. Essential items not fully covered in the company's provisional plans, e.g. handling facilities and extension of the bar mill, were also included. The consultants' support for the sheet mill element in the plan was a qualified one. Continuous casting of billets from the furnaces, a process then in its infancy, was considered but not recommended, although the consultants added, 'If, however, a further phase in increasing output should be envisaged, the question of continuous casting should again be considered.'

Although the board of ISH on the whole tended to play for safety as against experiment, they did not discourage management from investigating developments in steel-making technology which might be of benefit to the ISH plans. Before the consultants were appointed, George Radley was sent in November 1958 to inspect the pilot plant for continuous casting which one of the English steel companies had installed and in the same month the board directed that a suggestion by E. D. Bool, a former adviser to ISH, that electric-arc steel-making could be adopted for production of stainless or other special steels, should be followed up in reassessing the development project at Haulbowline. Apart from the advice tendered by their own consultants, the board were probably influenced by the adherence of the English steel-works to the conventional methods of steel production and rolling. Vaizey in his *History of British Steel* (p. 176) comments on the apparent lack of interest of the industry as a whole, even during the expansion of the early 1960s, in the adoption of new processes.

The cost of the plan could not be assessed without inviting

tenders for the major items of plant. These invitations were issued to a wide field of specialist suppliers, but only one comprehensive tender for the rolling mills and ancillaries was received. This tender was from Klockner of Duisburg, with sub-contractors designated from among the leading German plant manufacturers. The new open-hearth furnace, the melting shop equipment, the civil engineering works, the additional buildings required, the dockside cranes etc., were the subject of separate invitations to tender. The board had directed that as far as practicable, equipment, materials and labour should be procured in Ireland. While this condition was inapplicable to the main plant requirements, it was important in placing contracts for buildings, civil works and some items of handling equipment.

One item of expenditure not included in the John Miles assignment was the reconstruction of Spencer Jetty which had previously been considered when ISH were to be deprived of part of their dockside facilities by Verolme. O'Connell and Harley, the Cork engineering consultants later employed on the Haulbowline Bridge, had, at that time, drawn up plans for the Jetty. They remained in charge of the placing and supervision of the contract when it was decided to include these works in the development project. Barry O'Connell and his partner, Cyril Roche, undoubtedly displayed ingenuity and competence in all the tasks they undertook on behalf of ISH. Regrettably, both died at an early age.

The final plan was completed in October 1959 and submitted to the Minister for Industry and Commerce. It represented a considerable advance in scope and cost compared with the provisional scheme approved a year earlier and included in the government's First Programme for Economic Expansion at a figure of £2 million. The total estimated cost, based on tenders or, where tenders had not been received, on measurements by the consultants, had risen to £3.7 million including Spencer Jetty, road works, electrical services and other items not covered by the original projections. Outputs of ingots had been increased to 80,000 tons and of finished products to 64,000 tons annually. These were the projected outputs on which the viability of the scheme was based, but the plant had a reserve capacity to meet market opportunities beyond those already assumed. The company was able to make a modest contribution from its cash resources so that the capital investment required from the state was £3.6 million.

The mood of optimism and faith in the performance of reputable international contractors was undiminished when ISH discussed their plan with officials of the Department of Industry and Commerce, pressing on them the need for a rapid decision in view of the tendency for prices to rise above the tenders received. The main crux arose from the ISH contention that all of the state subscription should be by way of share capital in the company unless, preferably, part of it could be provided by free grant. The Finance view was against grant and in favour of a 50 per cent repayable loan; they were alarmed by the increased cost of the project and apprehensive of the risks of loss in a competitive marketing situation under conditions of free trade. The company did not press the claim for a grant but it was agreed that it should not be burdened with loan charges for interest and repayment. It was relevant to the issue that the company had survived profitably for over a dozen years without any payment from the Exchequer and to that extent had earned favourable consideration of full financing of the development by share capital. Finance were sceptical about the prospects of dividends on the shares and as events turned out, they were right, but not altogether for the reasons advanced by them in 1959.

The government approved the development proposals in early January 1960. No time was lost. The chairman and Farren visited London on 8 January to arrange a programme of action with the consultants who came to Haulbowline ten days later for further discussion and decisions. Letters of intent were issued to Klockner in respect of the main contract for the mills and to Maerz of Zurich for the open-hearth furnace, so that the manufacture of the plant could commence immediately. The first commitment was over £1.5 million and the development project was, at last, on its way.

The negotiation of contracts for all phases of the project occupied considerable time during 1960. By hard bargaining ISH had secured a price reduction by Klockner in their original comprehensive offer for supply of the mills and their ancillary equipment. This benefit was, however, rapidly eroded by price increases in materials and labour affecting other contracts for which firm offers did not exist when the overall cost had been assessed many months earlier. In all, some thirty separate contracts had to be placed, mostly on the Continent, with several important items going to Britain. The buildings contracts involving major structural steel went to Irish firms, Smith & Pearson of Dublin (melting shop and merchant mill), McGloughlins of Dublin (scrap yard) and Kelly of Portlaoise (warehouse). A key contract was that for the extensive work of demolition, excavation and new foundations in which efficient and punctual performance was a vital element in the whole project. The lowest tenderer on the basis of specifications formed by the consultants after some trial borings on the site by Cementation Ltd. was an Irish firm with Dutch associates, Ascon Ltd. of Dublin and Cork. Ascon were also given the contracts for Spencer Jetty and for a new item, the passenger landing pier, both completed on time and within the estimates.

The civil engineering contractors came on site in September 1960. They were there for over two years, the work being programmed in sections with one eye on the need to allow production to continue in the old plant and the other eye on the need to have each section of the foundations completed before the new plant arrived at Haulbowline for installation.

In October, two of the directors, O'Shea and Connor, accompanied by Radley, went to Kockner at Duisburg to arrange for the training of ISH operatives on German mills of the type being supplied and for further training and supervision by Klockner personnel during and after the commissioning of each unit.

For general monitoring of the progress of the plan, the ISH development committee remained in being. Farren was determined to maintain financial control, particularly when the first signs appeared that the estimates were wrong and he insisted that any changes in the plan, whether advanced by consultants or the contractors, which could or would involve additional expense or delay, must have his prior approval. This was an attitude which the executive director and the whole board fully endorsed, although the consultants must have felt irked at times and thought, no doubt, that ISH were trying to play too strong a hand in the details of the job. As problems and difficulties grew from mid-1961 onwards and differences between the company, the consultants and the contractors became manifold, joint meetings were held to attempt to resolve these. Those who participated in these deliberations (a few might justly be termed wrangles) can scarcely remember them with anything but pain! More than once it required all the placidity of which the chairman was capable to avoid an explosion louder than the detonations of Ascon's dynamite as it demolished or excavated nearby. The growing impatience and disillusionment of the directors was felt by all for, as no end appeared in sight, they began to question every promise and every assessment and forecast presented to them. The board minutes of the period contain many expressions of fretful concern and complaint. The principal actors in this drama (apart from ISH people) were, however, exceptional duellists and most worthy of attention, all very able men with diverse personalities and temperaments: George Brooke, a senior partner in John Miles; Rolf Doerter, a director of Klockner; and Pat Herlihy, the ever-smiling leader of Ascon.

Some of the controversies, particularly those in relation to the formidable civil works, arose from the company's determination to maintain production from the existing plant during the progress of development for as long as possible. The site became impossibly congested as a result. Contractors complained that they were not being given enough working space and time schedules suffered thus adding to the cost of the job. On the other hand ISH production staff blamed interference from contractors' activities for poor performance and consequent increased unit costs. The project had from the start been planned on the basis of minimal interruption of production. It was essential for the company to maintain its business, customers' goodwill and cash flow and to avoid the unemployment which any long close-down would entail.

Some idea of the congestion in the restricted site at that time may be gathered from the fact that during the progress of the works one of the contractors, Ascon, moved 100,000 tons of equipment and materials by boat from the mainland to the island, (the bridge, alas, was not yet there to avoid the costs of double-handling). This takes no account of the call on space as other contractors' materials, mill parts and diverse equipment began to arrive for installation. Facilities for the contractors' personnel had also to be provided. Pat Herlihy has stated in his published paper on the civil works (see chapter three), that in the case of the intensive merchant mill operation alone, Ascon employed over 300 men and six engineers on the site. Such problems are a part of any major development, no doubt, but seldom in such a limited and water-bound area and in the midst of a heavy industry which placed its production needs as the primary object of planning organisation and cooperation.

There was another sufferer from these development activities. This was the neighbour of the steel industry and the paramount occupier of the island. At the annual general meeting on 14 December 1961, the chairman expressed 'appreciation of the helpful attitude and cooperation of the Department of Defence and of the naval staff at Haulbowline which have materially lightened the difficulties of ourselves and our contractors in executing these large-scale works on our restricted site'.

The government decision that the cost of expanding the steel industry should be met from public funds required parliamentary approval. Specific legislation is the normal method of authorisation in the case of capital provisions such as the purchase of shares in a company where a series of transactions during more than one financial year is found necessary. As an interim measure, Jack Lynch as Minister for Industry and Commerce, sought a supplementary provision on his department's vote in March 1960 for a sum of £300,000. The enabling Dáil resolution was described as a Grant-in-Aid. The debate on the supplementary vote was short. Those who spoke (Sweetman, Larkin) supported the project.

Although the Irish Steel Holdings Limited Bill was formally introduced in Dáil Éireann on 13 July 1960, a second reading could not be envisaged until the autumn session due to pressure of more urgent legislation. In these circumstances Jack Lynch sought approval on 19 July for a further supplementary vote, this time for £1.2 million which, with the £300,000 already granted, would enable ISH to meets its capital commitments until the end of the year.

The Bill eventually came forward for a second reading on

27 October 1960. The main purpose of the measure was to authorise the Minister for Finance to buy shares up to a limit of £4 million and to sell or otherwise dispose of such shares. The opportunity was taken to include other provisions in the Bill, i.e. those relating to the giving of state guarantees of ISH borrowings up to a limit of £500,000, and to the requirements that the annual audited accounts of the company should be submitted to the Minister and presented by him to both Dáil and Seanad, and further that the form of these accounts should comply with any directions given by the Minister.

In his opening remarks Jack Lynch described the scope and cost of the development while stressing the national importance of the industry and the possibility that other engineering and steel-using industries would be established as a result of expansion at Haulbowline. Deputy Cosgrave accepted the Minister's approach and indeed praised ISH: 'The management of the company reflects credit on all concerned.' Recalling his own visit to the steel-works when parliamentary secretary in 1948, he specially welcomed the decision to complete and operate the sheet mill. Why should the public be asked to subscribe, was a Cosgrave question echoed by Deputy Esmonde, who also thought £4 million of public money was a great deal to invest in one industry. There was no doubt, however, about the welcome given to the Bill by Deputy Martin Corry, on the note 'Thank God I have lived to see the day', especially when the good news emerged that sheet steel was to be produced, a project for which the deputy had campaigned for many a long day. He was lavish in his congratulations to the government, to Sean Lemass, even to the management of ISH with special words of praise for the two general managers, first Fitzpatrick and since 1955 Farren. The Minister had little difficulty in replying to the debate. The Dáil was in equally acquiescent mood a week later when the Bill passed through Committee and final stages virtually without comment (*Dáil Debates,* vol. 184, col. 288 *et seq.*).

In October 1962 the company came to the conclusion that the limit of £4 million should be increased to £6 million and that the government should be so informed. If cost was one problem, there were others caused by the delay in execution of the whole project so that it overran the time planned, and the operational difficulties in the mills (but not the melting shop). In the case of the blooming mill (which with its ancillary equipment including the soaking pits was the most costly single item in the project) a fundamental change had become necessary. This large mill had been designed to roll ingots down to the billet sizes required for rolling in the merchant bar mill, but experience had convinced ISH that this method could not produce billets at consistent levels of output and quality. As a result it was found essential to adopt double-rolling, i.e. ingots to blooms in the large mill and blooms to billets in the section mill. Thus unit costs were higher than had been originally planned.

There were also problems in the merchant bar mill, while neither the wire rod unit nor the sheet mill had yet been commissioned. In the case of the merchant bar mill the contractors eventually agreed to correct the cooling bed at their expense but this work caused cessation of production from the mill for two months in 1964 with consequential loss of sales revenue. The wire rod plant was really an extension of the bar mill, a high speed unit producing the narrow gauge rods in coiled form. The market for small wire coils collapsed after 1962. Production was abandoned and the equipment was sold some years later to an English buyer.

The sheet mill had been installed as planned but the inadequacy of fresh water supply from Cobh (pending the completion of the local authority's new reservoir) postponed commissioning tests. In the meantime the board came to the conclusion that the project should be shelved. It had always been a doubtful starter. The ISH management, it should be said, had advised against it when the 1959 plans were being drawn. The final decision to abandon sheet rolling and to dispose of the mill was not taken until December 1965 when, the board minutes record (most unusually), Tom O'Shea dissented from his colleagues, no doubt feeling that the conclusion smacked of defeatism.

Sufficient reference has already been made to some unhappy conferences between ISH, their consultants and the contractors. These meetings normally took place at Haulbowline but in May 1962 the chairman, accompanied by O'Shea and E. A. Coleman, had visited Germany for discussions at Duisburg with Klockner and at Erlangen with Siemens, who had been entrusted with the contract for electrical equipment. That could be described as a happy occasion. The board minutes of 20 May 1964 record a decision that no further payments (mainly retention money at that stage) should be paid to the main contractors and that was the position until, at a conference lasting three days at Haulbowline in November 1967, the ISH chairman and Herr Doerter for Klockner reached agreement on the outstanding financial points. By that time all parties were sick of the affair and ISH were already engrossed in proposals which might make further progress possible on the basis of what had been achieved by the First Development. Nearly eight years had elapsed since the first contracts had been placed and all the plant supplied (except, of course the wire rod and sheet mills) had been in use for several years. It is only fair to add that John Miles and Partners had never ceased, during the long haul to final settlement, to give their best professional services in circumstances of unusual difficulty. Throughout the protracted vicissitudes, with so much scope for disagreement or irritaion, ISH found them at all times honourable and agreeable associates.

It would be a mistake to conclude from this recital of misadventures that the First Development failed in its primary objectives. When Sean Lemass, as Taoiseach, honoured the works with a visit on 18 April 1963, he was obviously impressed by the achievement. He remarked that any resemblance between the plant he had last seen in 1957 and that now installed was, in his words 'purely coincidental in minor respects'. It was some consolation too that when Heffernan and McPherson, the Canadian experts, visited Haulbowline they observed that the plant taken as a whole was good value for the money expended and that in their experience any new rolling mills 'needed to be kicked around' for quite a while before their potential could be realised. Jack Lynch also inspected the development early in 1963. The succinct explanation of the excess cost given by ISH in a contemporary memorandum for him and putting the best face upon the matter was 'revisions of tender prices and unexpected difficulties in installation of plant, particularly in relation to the civil engineering works'.

Some reference is needed to the Irish Steel Holdings Ltd. (Amendment) Bill 1963 which proposed to increase to £6 million the limit within which the Minister for Finance could purchase shares in the company. Jack Lynch moved the second reading of the amending Bill on 27 March 1963. He told the Dáil he had considered curtailment of the development scheme but had found that to do so would nullify its purpose. He was satisfied from a personal visit to the works that the increased costs had been inescapable. As to the provision of £6 million, the Minister explained that this sum included a margin of £750,000, above the then known requirements of the First Development, which could be applied later to capital purposes necessary to equip the industry for free trade conditions (*Dáil Debates,* vol. 201, col. 475 *et seq.*).

Deputy Cosgrave was critical of the excessively high cost of the foundations at Haulbowline. Deputy Esmonde again took the line that ISH was being pampered in preference to equally deserving industries, and he suggested some degree of political motivation. The Minister firmly rejected this suggestion (*Dáil Debates,* ibid, col. 483). The leader of the Labour Party, Deputy Norton, who as Minister for Industry and Commerce (1954–57) had encouraged the steel industry, spoke about the difficult international situation for a small industry, with the added complications of Common Market conditions and increased competition. He raised the question of whether the industry would survive. 'It will be hard going,' he said, 'and those who can make Haulbowline a viable enterprise in the circumstances of today and the circumstances of tomorrow will deserve some form of decoration in due course from the Irish people.' The Minister replying to the debate dealt with the question of survival 'in free trade conditions'. It was too early to form judgments. ISH, themselves, he said, could not do so because they were still uncertain as to the terms of entry into the European Community. At that moment ISH selling prices were lower than the highest home prices in the Community, but higher than the lowest. Jack Lynch did not get as rough a passage as he may have feared. The Bill was passed through all its stages on the same day without a division.

In Seanad Éireann, the amending Bill was passed through all

its stages on 3 April 1963. The discussion reflected the criticisms and doubts expressed earlier in the Dáil. The Minister rejected suggestions by Senators Hayes and Fitzpatrick that there might have been costly mistakes in planning, chiefly as regards the foundations, and that the nomination of directors who were not businessmen might have been unwise (*Seanad Debates* vol. 56, col. 699/714).

The issued share capital has not (1978) reached the limit provided in 1963. Since 1967, after the completion of Haulbowline Bridge, it has stood unchanged at £5.785 million. The cost of the bridge, and also £250,000 representing a special provision made by the Minister towards dead-weight expenses falling on ISH during the development period, may be deducted in comparing the actual cost of the First Development with the sum of £3.6 million which the state originally undertook to meet. On that basis, the final cost is shown as approximately £5.2 million involving an excess of 44 per cent. Of that excess over the 1959 estimates nearly one half was incurred under Ascon's main civil engineering contract for reasons which were well established.

Notwithstanding the abnormal foundation costs, the expenditure on development could not be regarded as excessive. In the early 1960s the index normally used as a rough guide to the probable cost of erecting a steel works was £100 for each ingot ton of planned production. Admittedly this standard of probable cost applied to large fully integrated works. After the development expenditure, which virtually created a new industry, ISH was producing 75,000 tons of ingots each year until furnace capacity was further increased in 1972. Cost of equipment and expenses of installation undoubtedly rose subsequent to 1958–59 and estimates made in that period were inevitably exceeded. That was certainly the contemporary experience in Britain as manifested in the case of the two large steel works built at Llanwern and Ravenscraig where the planners' estimates of capital requirements proved seriously inadequate. One compares the circumstances of these great ventures with the relatively minor operation at Haulbowline solely to reinforce the point that the ISH excess expenditure did not necessarily represent bad estimating or poor execution, nor did the amount involved, although substantial, impose a burden greater than the Irish industry could bear.

It remains to give some account of the expenditure incurred in building Haulbowline Bridge. The estimate of cost approved in February 1964 was £300,000, based on Cyril Roche's admirable design and quotations for the work put to tender. The eventual total cost was about £320,000 of which £240,000 was paid to Ascon under their contract. The balance was for preparatory work undertaken by ISH, the cost of reinforcing steel supplied by ISH and the consultants' fees. Considering the length and strength of the two-lane road bridge and the time taken from contract to completion, over two years, during which wage and material costs rose, the performance was very satisfactory. It was a pleasant contrast to some of the earlier experiences during the course of development.

If a judgment on the First Development is to be made after nearly twenty years, one is tempted to ponder on the question whether the old saying 'too little and too late' may be applicable. Taking first the question of timing, it is probably true that a similar project conceived by ISH and approved by the government five years earlier could have been erected at significantly less cost. The export opportunities available during 1957–62 might have been met in a greater degree and more efficiently from an improved and extended plant. Apart from exports a suitable development completed in the late 1950s would have put ISH in a much better position to satisfy the expanding home market during the following decade. There would have been ten years instead of five to make profitable use of the new plant before further capital investment became necessary to gear the industry to technological changes and to the increasingly prevalent conditions of less protection and more competition. To some extent these considerations represent wishful thinking because, in reality, neither the directors nor the government were prepared for a major state investment in the steel industry as early as, let us say, 1954. It was only after Sean Lemass returned to office in 1957 and later when the first Programme of Economic Expansion was being formulated by the Department of Finance that costly changes at Haulbowline became an accepted part of policy.

The same consciousness that the steel industry was low

in the order of priorities for state investment from limited resources influenced the board of ISH in their approach to the development scheme. The lower the cost, the more likely would be government approval. Hence every effort was made both by ISH themselves, and by their consultants as directed, to make use of existing equipment or parts instead of, in all cases, seeking a completely new unit. For instance, with hindsight it would have seemed wiser to scrap completely the existing merchant bar mill and install a completely new modern mill of the latest design and of greater range at a cost in excess of the extensive improvements to the existing mill adopted in the development scheme.

For the same general motives of cutting expenditure to the bone, the directors shelved the project for a road bridge. This was a mistake. Again, with hindsight, the construction of the bridge might well have been given priority and, if possible, completed before the rest of the development was undertaken.

The case of the sheet mill has already been discussed. It had been a sort of 'King Charles's head' in all the company's thinking and planning for expansion and it was difficult to cut off. A more ruthless view might have suggested the exclusion of sheet rolling as a future possibility once the proposals for joint venture with RTB had come to nought in 1955. Had that been done the money allocated to the sheet mill in the development project could have been diverted to capital purposes of a more promising nature. The welcome given to the sheet mill project in the Dáil debate of October 1960 illustrates how difficult such a decision would have been.

One cannot blame the directors and management of ISH for not embarking on continuous casting of billets from the furnaces. In 1959 it was still an unproven process for all types of steel making or at least had not gained general acceptability. If there is regret that the opportunity was not taken by the adoption of continuous casting to avoid the heavy investment in the blooming mill and soaking pits, this merely creates the suggestion that the ISH First Development was undertaken too early rather than too late. The Irish steel industry could not delay expansion. And how long to wait? Even twelve years after 1959 ISH had much difficulty in persuading the Electricity Supply Board to accept its first electric arc furnace. (The advice given by the consultants in 1959 was that continuous casting would prove more effective in conjunction with electric furnaces which heat at short intervals than with open-hearth furnaces .) In 1959 steel production by open-hearth furnaces was still the process in most use throughout the world.

There was another factor which in 1959 restricted choices in the planning of development. The state programme of investment in industry demanded that new capital should create jobs. Steel was, however, a capital intensive industry, in which technical improvements were directed towards achieving greater plant outputs while reducing man-power ratios by mechanical and electronic devices. ISH were not seen to be a preferred object of state investment because increased employment could not be generated on a scale commensurate with the amount of public funds involved. The First Development envisaged an increase in work force of only from 500 to 700 after £4 million, later increased to £6 million, had been expended. It was a point of criticism in government circles. There was, thus, an inclination in ISH to avoid marginal expedients which would reduce employment. This caused a degree of over-manning in relation to output, which, as wage levels rose, became an embarrassment to the industry.

What should the verdict be? Probably the answer is that the First Development was neither a huge success nor a depressing failure. It fell between these poles, but the gains outweighed the disappointments; not indeed a model performance in planning or execution but a considerable achievement in the contemporary setting. The management of ISH learned the hard way lessons which were to serve them in their approach to future proposals for expansion and in their selection of the means to implement them. The industry, itself, was far stronger than it had been and better equipped to meet the challenges of the future. It was a first step, but a long one.

CHAPTER X

ISH Comes of Age but does not Wed

This chapter deals with the five years immediately following the eventual completion of the long drawn-out process of development culminating in the opening of Haulbowline Bridge and the installation of the second galvanising line. The chapter heading indicates that the period included an event of some interest to the company but of little extra-mural significance and a courtship that became a non-event because the prospective bridegroom changed his mind. ISH is placed in the female role in that affair as clearly the weaker party whose dowry failed to match the weightier financial pressures on her suitor from another quarter. However, although no marriage resulted the parties remained on speaking terms. Their relations had a bearing on the subsequent decision by ISH to proceed to what was called the Interim Development, a project which demanded intense efforts during the latter years of the period chosen for review.

When reflecting on the 'coming of age', the directors were not so much concerned with the institutional milestone as they were with the long service of those who had continuously worked for the company since its birth in 1947. It was considered fitting to mark their survival with some contrived rejoicing. Gold tie-pins were designed embodying the company's crest and a suitable inscription. These tokens were intended to be a service decoration of honour and not in any sense material reward. George Colley, Minister for Industry and Commerce, visited the works on 21 October 1968 and in the presence of a general meeting of all employees presented the tokens to the forty-one individuals entitled to receive them. For good measure, the two original directors, O'Shea and Hogan, were included among the recipients. That evening the 'long-service men' were guests of honour at a dinner given by the company at a Cork hotel. The idea of thus honouring completion of twenty-one years service was found so acceptable that it was decided to continue it. Since the first event in 1968 each year has produced a new batch of eligibles for the ISH gold pins, although fewer in number than the original group of survivors from 1947. Representatives of the various works committees and associations are invited to attend the dinner. These occasions are now established as a part of the traditions of ISH. The celebrations give pleasure and pride to those who have earned the long-service tribute. They are also good for morale and stimulate the family feeling that is so necessary in any industry.

When the ISH report and accounts for the twenty-first financial year were presented to the annual general meeting at Haulbowline on 30 October 1968, they were seen to merit some blowing of trumpets. There was a net profit of £433,000 after meeting all charges on a turnover of over £4 million which included export sales of £550,000. The balance sheet was healthy, no bank overdraft, cash, including deposit funds, of £344,000 and liquid assets exceeding current liabilities by nearly £2 million.

In his statement at the meeting the chairman commented:

> These results were achieved not by luck or in favourable conditions but by the determination and ability of management and workers to overcome the difficulties caused by rising costs, falling protection and fierce competition in external markets. They were achieved, too, without any increase in the company's selling prices in Ireland.

In fact there had been no increase in selling prices since 1964. This was somewhat remarkable during a period when the national price index continued to rise. It was, however, consistent with the board's oft-enunciated policy of, so far as practicable, maintaining stability in their domestic selling prices and at a point below the level allowed by the umbrella of protective duties. Practicability, of course, depended on factors of cost outside the company's control. Whenever the policy could be applied, it clearly benefited Irish consumers of steel, chiefly in the construction industry. In a few years it was no longer possible for ISH to maintain price stability as steel production, not only in Ireland, became overwhelmed by the rising tide of inflation. In these later circumstances did the ISH directors sometimes regret that they had not, when times were favourable, moved selling prices upwards and, by adding to the reserves, increased the company's financial strength and ability to face the difficult years? It is more probable that even if the directors had been able to foresee the future, they would still have adhered to their policy of price stability and moderate profits, knowing that a quest for high profits would have been unacceptable in national economic terms and would have accelerated the inflationary pressures within the industry itself.

Shortly before the 1969 annual general meeting the company broke new ground by holding a press conference to give information about the industry and its future plans. The journalists visited the works and inspected the production processes. At the press conference there were several questions about a document which had recently been given publicity by the newspapers, having been circulated to Ministers and public representatives. The document, in the form of a complaint requesting action by the Restrictive Practices Commission, was signed by individuals named Flavin, Eastwood and Doody. The complaint assailed the ISH scrap purchasing practices as being discriminatory (in favour of Hammond Lane Metal Company) and contrary to the public interest. Flavin was a former managing director of the Hammond Lane Metal Company; Eastwood was a leading Belfast scrap merchant and Doody was a chartered accountant in Dublin with no known connection with the scrap trade. It was an odd feature that these people had never approached ISH about their complaints nor had a copy of their statement been sent to the company. At the press conference, the chairman replied to the questions by stating that the company had not received a copy of the document but that from the extracts published by the newspapers, it appeared to contain many false and misleading statements and inferences. The full significance of this is discussed below.

As a sequel to the interchange of visits and subsequent discussions recounted in chapter eight, a team of GKN technical experts headed by M. P. Rathbone was deployed at Haulbowline. GKN furnished their report, termed a 'feasibility study', on 27 January 1967. This survey contained short-term proposals which, without increasing plant capacity, would, at relatively minor cost, improve outputs and profitability and also long-term proposals for raising steel production to 250,000 tons a year involving an estimated capital outlay of about £6.5 million including working capital. If it were quickly decided to adopt the major scheme, the short-term proposals would become irrelevant.

An interesting observation in the GKN survey of the existing industry, in the light of the many headaches during the First Development, was that the mills had generally been 'well designed and properly installed' and that the blooming and section mills 'formed a highly flexible combination capable of high outputs'. The core of the long-term plan was that these mills should be used exclusively for rolling alloy and carbon forging quality billets and rounds for export. The special quality 2-ton ingots to feed the rolling mills, up to an annual tonnage of 250,000 tons, would be produced from a battery of four 70-ton Maerz open-hearth furnaces (of which one already existed) working on a cold charge of 60 per cent pig and 40 per cent scrap. This arrangement would leave the merchant bar mill entirely dependent on imported billets to be rolled to bars and angles required by the Irish home trade or for such exports as might be feasible up to an annual capacity of about 70,000 tons. The galvanising plant would continue to be operated on the existing lines. For production of quality billets and rounds at the projected level of output considerable new ancillary equipment would be required to serve both the melting shop and the mills. After completion of the scheme in two years, the GKN study envisaged a profit of 15 per cent per annum on the total assets employed in the industry of about £12 million (including the £5/£6 million of existing assets).

The GKN recommendation in favour of additional cold-charged open-hearth furnaces for steel-making as 'technically desirable, cheaper in capital cost and lower in running costs' had not been reached without some consideration of alternatives. The low tonnages projected and the lack of native raw materials were held to preclude production of hot metal by conventional methods at economic prices. Production of the required additional steel by electric arc furnaces of commercial size, 'the most attractive method all things being equal', would not be tolerated by the ESB, because of the flicker and other interference caused by the continuous operation of such furnaces with heavy loading on the electricity system serving the area in which Haulbowline is located.

The feasibility study was examined closely by ISH management who had many questions to raise. Their views were considered by the board who felt that the long-term plan offered sufficient promise to justify a request to GKN to define their proposals in greater detail chiefly as regards material costs, manning, marketing and financial projections. The board decision to move forward was taken at their meeting of 21 February, an occasion when they noted with deep regret the death of their solicitor. Ever since the formation of the company in 1947, Dermot McDowell, a most astute lawyer, had been the friend and confidant of the directors as well as their legal adviser. He was succeeded as principal of the firm, John M. McDowell and Company, by his son Denis, who continued his father's role in the affairs of ISH.

The GKN consultancy team again came to Haulbowline and, with the assistance of the ISH staff and by closer investigation of procurement and installation costs, production targets and manning, sought a realistic projection of profitability as well as an assessment of methods of financing the investment. These operations occupied some months. It was not until late in May 1967 that the second GKN feasibility study was received. It confirmed the major recommendations, and the financial projections were if anything more encouraging than in the preliminary report. One difficulty for ISH, however, was that the large investment would not generate any significant increase in the numbers employed in the industry.

At this stage it had become a possibility that GKN might be interested in the scheme in other ways than as consultancy advisers. In talks with Bruce Gardner ISH had expressed the view that, for success, the recommended project required active assistance from GKN in the early stages of production and in marketing the quality steels. Participation was not then discussed but it was apparent that this subject might become opportune as a result of events in the British steel industry. The issue of nationalisation of that industry had been a football in British politics since the war. Partially nationalised by the Labour government in 1950, partially re-vested in private ownership after the Conservatives returned to power in 1953, the fate of the industry was finally settled by the Labour successes in the elections of 1964 and 1966. The Iron and Steel Act of 1967, effective from 28 July of that year, transferred all the major crude steel producing works to public ownership under the new British Steel Corporation (BSC). Among those transferred undertakings were the large GKN works at Scunthorpe and Cardiff. The fate of the smaller Brymbo works was still uncertain, but its eventual transfer to the BSC appeared likely. GKN was the largest engineering industry in Britain, the main supplier to the UK motor vehicle producers, drawing its steel requirements from the works it owned and operated. In the new situation, GKN would become dependent for its steel on the BSC unless it purchased abroad. Raymond Brook, the chairman and chief executive of GKN, had never concealed his opposition to nationalisation of the steel industry.

The chairman of ISH was not greatly surprised, therefore, when towards the end of June 1967 Brook telephoned him saying he had a proposition to put to the Irish authorities, and would like to present it at a private meeting if that were possible. He came on 3 July to Haulbowline accompanied by Bruce Gardner and two other GKN directors, toured the works and discussed his ideas with the chairman and Farren. In Dublin on the following day, during an informal lunch, which Jack Lynch and the two Ministers concerned with ISH affairs, Charles Haughey (Finance) and George Colley (Industry and Commerce) attended as well as their permanent secretaries, T. K. Whitaker and J. C. B. MacCarthy, Brook fully and candidly explained his company's interest in using Haulbowline as a source of steel supplies to GKN, provided control of an expanded industry by GKN on mutually agreed terms was not repugnant to Irish government policy. The Taoiseach was, naturally, non-committal. It was, however, suggested that a tentative scheme be explored between the departmental and company representatives. If agreed proposals emerged they could be submitted for consideration by Ministers. Later that day in the Department of Finance,

Whitaker, MacCarthy, Hogan, Farren and the four GKN directors discussed in outline the possibilities of a partnership arrangement involving financial and commercial undertakings by both parties. Considerable progress was made without a definite conclusion being reached.

Writing to the chairman of ISH after his return to Birmingham, Brook expressed some disappointment. 'Possibly', he wrote, 'each of us expected more from the other than it was possible to concede,' but he was hopeful that solutions could be found at a subsequent meeting. In the meantime everything was to be treated by both sides as a confidential topic to which the label 'Project Ivor' would be applied in all subsequent correspondence internally and externally.

At a further meeting in Dublin, when GKN were represented by James Insch, the second-in-command, and by Rowbottom, their financial director, substantial accord was reached on the framework of an agreement which could be further considered by both sides when a formal draft was ready. It was not until October 1967 that word came by telephone from Brook to Haulbowline reassuring the chairman of GKN's continued interest and repudiating newspaper reports that GKN were involved with American companies in plans for a major steel works at Bantry Bay. As regards GKN'S decision to undertake in partnership with BHP Steel Company the erection of a major integrated works in Australia (a matter already public), Brook said that this development would not prevent GKN from embarking on the relatively minor investment in Ireland if proposals on that issue should be found mutually acceptable. The ISH chairman replied that the delay was embarrassing, adding, 'We are not without other suitors and unless you offer the ring soon, we shall have to break off the engagement.'

In January 1968 the draft of an agreement for giving effect to Project Ivor was discussed at GKN head office with Insch, Bruce Gardner and Rowbottom. GKN confirmed that subject to some minor qualifications which they proposed, the provisions in the draft correctly represented the outcome of the Dublin talks. There could be no decision, however, until Brook had put the question to his full board. It appeared that the existence of Project Ivor had been leaked to the British government with unfortunate consequences for GKN.

The ISH chairman made it clear that the proposal had been kept off the record in Ireland. Project Ivor was treated as a 'non-event' in official quarters pending agreed proposals. GKN had offered at the first meeting (as was now explicit in the draft agreement) to undertake permanent major investment in a reconstructed Irish steel company which would sell most of its output to GKN. If this were not still the GKN position, the chairman stated, the whole project would fall to the ground. Writing to George Colley after his return from Birmingham, the ISH chairman expressed doubt about the outcome but thought it would be worth waiting a bit longer.

The wait proved longer than the three months Insch had asked for in January. It was not until July 1968 that Brook and Insch came to Dublin on what must have been in the circumstances an uncomfortable mission. Brook told the ISH directors that it was no longer expedient for GKN to proceed with Project Ivor, the mainspring of which was the annual sale by ISH of up to a quarter of a million tons of forging quality steels to the GKN group of companies, because the British Steel Corporation had offered such favourable terms for a commitment by GKN to purchase virtually the whole of its steel requirements from BSC works that, apart from policy, it would be financially impossible for the GKN directors to reject these terms. Brook disclosed, in confidence to Hogan, information which seemed to prove the financial point. He regretted the abandonment of the Irish venture, to which he personally had been greatly attracted, and the inconvenience caused to the ISH board and their authorities by the GKN withdrawal after so much delay. While investment in ISH on the lines of Project Ivor was no longer practicable, the GKN chairman said he would be glad to consider some alternative method of assisting in the development of the Irish steel industry.

Thus ended the courtship of ISH by GKN. The 'first fine careless rapture' had evaporated on both sides long before Raymond Brook's withdrawal. An unhappy marriage may possibly have been averted by the failure to reach full agreement on the first urgent approach in July 1967. Alternatively there might have been an engagement eventually broken under political duress, a sensitive and potentially inhibiting factor on both sides.

The ISH board and management may indeed have had some feelings of relief that the uncertainties of Project Ivor had ended since they had much to do during 1968 and 1969 in order to maintain the momentum of improvement in production and trading. With their limited steel production, the rolling programmes had to be rationalised to serve those products in highest domestic demand, while export targets were necessarily reduced. Flats and joists were excluded from the production range and Northern Ireland was given priority for exports of such surpluses as could be spared from the buoyant home market. Long-term contracts for Eastern European pig iron were arranged, thus providing the unusual spectacle of freighters flying the Soviet flag discharging cargo at Spencer Jetty. It was a case of buying suitable material in the cheapest market.

Increased wage settlements eventually forced the board to raise their domestic sales prices. The minutes of 20 August 1969 record the decision to ask the Prices Commission to approve an increase of £3 per ton for all rolled products. The selling price of galvanised sheets was undisturbed for the time being, but it too was raised by 10 per cent early in 1970. They were the first increases since 1964.

The outcome of the company's financial year ending 30 June 1969 was reported to the annual meeting on 12 November. There had been satisfactory increases in outputs (21 per cent) and turnover (29 per cent), with exports valued at £650,000. The net profit was returned at £525,000, but the chairman warned that this level of profitability could not be maintained in face of rising costs (£10 per ton since 1967 and still rising) and the limited capacity of the existing plant. The company intended that this latter difficulty would be remedied by a substantial addition to steel-making facilities as an interim development during the next two or three years in preparation for a later major expansion in the industry. One encouraging feature of the report was that there had been during the year an increase of 18 per cent in productivity from the numbers employed in the industry – now risen to 832 – althought this was not enough to compensate for the effect of wage adjustments on an admittedly high level of manning in a small steel works.

These accounts were certified for the last time by the

comptroller and auditor-general who, it will be recalled, had been appointed company auditor at the outset in 1947. The official auditors who came each year to Haulbowline had always been most welcome and helpful. It was no reflection on them that the ISH board concluded that the time had come for a change. The Minister for Finance, Charles Haughey, accepted the chairman's request that ISH should be allowed to appoint commercial auditors as being more suitable to a company operating in the competitive market place and conducting international business. There were, of course, precedents for the employment of commercial auditors by state bodies, as in the case of the Air Transport Companies and the Electricity Supply Board (ESB). In face of some umbrage and contention from the auditor-general, the Minister abided by his decision. He told the chairman he could choose any auditors he wished, 'provided they are not the firm with which my name is associated'. The choice fell on Magennis, Burns, Griffin and Company of Cork (later reconstructed as Cooper Magennis) who agreed to accept the nomination. In accordance with the ISH Articles of Association, they were appointed at the general meeting in November 1969 and have been retained in office from year to year since then.

The board minutes of 1969 record a decision to make an ex gratia contribution of £10,000 towards the improvement of the country road leading to the approach to Haulbowline Bridge. This followed from representations by the County Manager in a letter to the chairman pleading financial reasons for delay in carrying out these necessary works. The board were not averse to appearing as benefactors of the local authority, particularly in a case where public spirit and self-interest coincided. In addition to the grant they transferred one and a half acres of ISH property on the mainland to the County Council for a nominal consideration. This transaction (like the contemporary donations for community purposes in Cobh) provides some evidence of the steel company's growing financial confidence and strength and its acceptance of a leading role in the industrial life of the area. A few years earlier a more austere view would have been taken.

The main preoccupation through 1969 and 1970, at least as regards expansion, was how best to increase rapidly and at the lowest cost the output of ingots from the melting shop. The target was 150,000 tons annually, nearly doubling capacity, and if the ESB objections could be met, this would be achieved by a thirty-ton electric arc furnace. Soaking pits would be extended and also the merchant mill facilities, involving civil works at the southern end of the works. The whole project, on which GKN consultancy and design services were engaged, with O'Connell and Harley associated on the civil works, eventually emerged as one involving capital outlay of about £3 million. In July 1970, the board submitted the plan to the government on the basis that ISH would meet one-third of the cost from its reserves and cash flow, the remainder to be borrowed by the company on five/seven year repayment terms at low interest available in the UK under the export credit guarantees available where British contractors were used. The IDA were kept informed. While not a pre-condition of the project there existed hopes that in due course an IDA grant might be forthcoming. The following October the government approved of ISH entering into commitments to implement the plan, henceforth to be known as 'the interim development'. The Minister for Finance agreed that, if required to do so in order to secure the most favourable loan terms, he would guarantee the company's borrowings in London. The Irish Steel Holdings Limited (Amendment) Act 1971 increased from £1 million to £3 million the statutory limit on guarantees of ISH borrowings.

At a press conference and lunch in Dublin on 16 November, the chairman announced the company's interim plans and gave some information about the company's results for 1969–70, which were reported to the annual general meeting at Haulbowline two days later.

The general manager, Farren, had unfortunately fallen ill. He recovered early in the new year but had a relapse and did not resume duty until the beginning of March 1971. In his absence E. A. Coleman acted as deputy and with the chief accountant, Jack Adair, completed the negotiations for a loan of £2.2 million from a London banking consortium headed by Morgan Grenfell and Company.

The main contract for the interim development was entrusted by ISH to GKN (Contractors) Ltd., the company in that group which undertook major installation contracts for steel and engineering industries in Britain and abroad. The ISH board were satisfied as to both the competence of this contracting organisation and the GKN commitment to making a success of the Haulbowline assignment. After tenders, major sub-contracts went to British firms for the arc furnace and for the costly compensatory equipment required to reduce possible disturbance of the electricity system to a level tolerable by the ESB, and to the Cork construction firm, HMC (Hegartys) for the civil engineering works. It was hoped that the whole scheme would be implemented within fifteen/eighteen months from the placing of contracts, with minimal interference with existing production.

In his statement to the annual general meeting in November, 1970, the ISH chairman expressed concern about the 'gross inflationary pressures to which our wage and material costs are subjected at the present time'. Net profit had declined to £442,000 from the record figure the previous year. None the less, the company balance sheet remained strong with a high degree of liquidity.

An unanticipated financial difficulty had arisen from the prolonged closure of the Irish commercial banks due to strike action by their employees during the summer and autumn of 1970. To meet their current obligations during the closure, including foreign payments to suppliers, ISH obtained facilities from a licensed banking institution which remained open, United Dominions Trust (Ireland) Ltd. (UDT). They provided cash for wages and transacted ISH foreign currency commitments. In this case and when there was another bank strike a few years later, it was fortunate that alternative banking facilities enabled essential payments to be made. However, there was no way, while the Irish clearing banks remained closed, of bringing to credit customers' cheques drawn on these banks. The bank strikes were symptoms of the industrial unrest caused by inflation, when strongly posted sectors of unionised labour sought to enforce claims to higher wage settlements than those allowed by the National Wage Agreements. ISH themselves were subject to such pressures on an increasing scale from 1970 onwards, particularly from the craft workers, who claimed that their differentials had been eroded and should be adjusted under the 'anomaly clauses' of the national agreements.

Wage claims and other such difficult and abrasive subjects were primarily matters for the personnel manager Harry Kerr (who had been promoted from the engineering department in May 1969). In February 1972, the board decided that the delegated powers of the executive director included the settlement of all questions concerning recruitment, pay and conditions of employment and that the board as a whole should be consulted only in relation to a matter involving some new major principle or having a substantial effect on the company's financial position. The executive director and his management team continued to resist all demands for wage concessions beyond the agreed national settlements. The arguments for wage restraint were coercive in national economic terms, but apart from that the steel industry itself was unable to absorb higher costs. Patience and adroitness, qualities fortunately possessed by the personnel manager, averted collision courses (until the disaster of 1977) although bans on overtime and threats of strike action were not uncommon. A serious symptom of industrial unrest was the decline in productivity.

The situation of ISH at the end of 1971 was described when the report and accounts for the previous financial year came before the general meeting on 8 December. As anticipated, net profit had again fallen, to £366,000 from £442,000, and the decline continued. There was concern at the fall in domestic demand caused by a general economic recession, a reflection in Ireland of international conditions. The world steel industry had begun one of its cyclical depressions, with supply greatly exceeding demand. This made Irish steel exports increasingly difficult to achieve. In reference to the probable accession of Ireland to the European Community by 1973, the chairman stressed the vulnerability of the steel industry unless an adequate transitional period and reasonable dispensations were allowed. Repudiating the prevalent idea that joining the Community meant what was termed 'free trade' he went on to say:

> Viewed internationally there is no such thing as free trade in the steel industry. In the Community itself there are regulations and restrictions and protection against non-member countries. Provided no unfair trading is practised

against the Irish steel industry and provided the government can use its powers to defeat dumping, this company faces new conditions and the obvious problems of adjustment with reasonable confidence.

Over the past twenty-five years a strong steel-making community has been established around the lower Cork harbour. Fathers and sons work together in our industry. The livelihood of many hundreds of families is at issue. It is impossible to believe that the ability and determination to succeed which have created a deep-rooted industry will not be equal to the challenge of the future.

Professor Eamon Dillon was co-opted on to the board of directors early in 1972. He was (and still is) the head of the Civil Engineering department of the Cork University. His professional attainments and wide experience, which included exciting events in 1944 during the preparation and installation of Mulberry Harbour for the Allied landings in Normandy, as well as his keen interest in the physical problems of the industry, made the professor a most valuable and congenial addition to the board. It could be said that 1972 was on the whole an unhappy year for the company. Extreme difficulties arose in almost every phase of production and marketing. The home market declined further owing to the continued depression in the building and construction industry, although there was partial compensation from improved sales of galvanised sheet. In order to dispose of surplus steel production and thus maintain employment in the works, Coleman was despatched to Britain with instructions from the board to obtain orders for 10,000 tons even at exigent prices. He accomplished his mission quickly but the troubles from these unsatisfactory orders persisted well into 1973. Material and wage costs rose steadily but it was not easy to get timely approval from the Prices Commission for increases in ISH selling prices.

The interim development was completed in reasonable conformity to the time schedule but its cost increased about 20 per cent above the original estimate of £3 million. This was unavoidable in a period of rapidly increasing prices, particularly as the high cost of the compensator (circuit breaker) equipment, a truly formidable installation, had been underestimated. More serious than cost were the difficulties and setbacks experienced when some items of the new plant came into service. In February 1973 an explosion and fire in the circuit breaker put the arc furnace out of commission for several weeks and raised questions between ISH and the contractors as to the design and security of the installation.

All these untoward happenings in 1972 caused financial difficulties. It was necessary to ask the Bank of Ireland to increase the overdraft limit to £600,000, while in addition, UDT agreed to make advances up to £500,000. The company's cash position was, however, greatly eased when the Industrial Development Authority decided early in 1973 to make a grant of £800,000 towards the cost of the interim development.

The Minister for Industry and Commerce, Paddy Lalor, had decided in April 1971 to refer to the Restrictive Practices Commission the complaints first published in November 1969 about ISH methods of purchasing scrap. The Commission undertook to hold a public sworn enquiry but the hearing did not begin until 1 May 1972. Prior to that the company had submitted to the Commission a full statement of their purchasing policies and methods, and of the necessity for them if the steel industry was to receive adequate and continuous scrap supplies. At the hearings, which continued until mid-September, ISH were represented by N. McCarthy S.C. and A. Murphy B.L. (instructed by the company's solicitor). The Commission visited Haulbowline during their investigation so as to inspect the ISH methods of receiving, grading and weighing scrap which had been criticised in evidence given by the complainants. At the public hearings, evidence on behalf of ISH was given by the general manager, by E. A. Coleman (until his tragic death on 18 June) and by John Rafferty. Hammond Lane Metal Company and the complainants were also represented by Counsel and submitted evidence. The main burden of complaint was that the ISH purchasing methods (which had existed for over twenty years without previous challenge) created, whether intentionally or not, a monopoly for Hammond Lane and that this situation had been aggravated by D. D. Frame's presence on the ISH board since 1967. On the latter point, it was established by evidence before the Commission that while a director of ISH, D. D. Frame had not taken part in negotiations between Hammond Lane and ISH concerning scrap supplies. In fact, scrap purchasing was a routine matter for management and not one for board decision. The executive director dealt with any special problems that arose.

It was a great surprise to ISH when in December 1972 the Commission reported in terms which were clearly adverse to ISH, and more receptive to the complaints of unfair trading than seemed to ISH to be warranted by the circumstances. The Commission recommended to the Minister for Industry and Commerce that he should require ISH to adopt a new purchasing system which would abolish the company's system of prices graded upwards according to the tonnage received from the supplier, and instead pay the same price per ton to every supplier who sold a minimum tonnage each year to the steel works. This system, it was claimed, would encourage small scrap merchants to enter the business and sell directly to ISH instead of through Hammond Lane. The Commission also recommended that the Minister should remove David Frame from the ISH board.

When asked by the Minister for Industry and Commerce to comment on the Commission's report, the board at once re-affirmed their confidence in David Frame's integrity and in his value to the company as a director, and urged the Minister to reject that part of the Commission's recommendations. On the main issue of purchasing methods, the board stated their belief that under the system recommended by the Commission, ISH would obtain less scrap and have to pay more for it and that scrap quality would decline if reliance had to be placed on direct supplies by small dealers, who would not have the capital to invest in processing equipment. The ISH directors considered that the report did not adequately present the case made by ISH. The petitioners' memorandum of complaint was published in full in the report but the statement submitted to the Commission by ISH rebutting these complaints was excluded.

The Minister for Industry and Commerce did not accept the Commission's recommendation that he should remove one of the ISH directors. Paddy Lalor confined himself to directing that ISH should implement the purchasing terms recommended by the Commission and this was done. In the annual report in December 1973, the company stated that since the new terms had been adopted, domestic supplies of scrap to the steel works had slackened and the number of direct suppliers had fallen. It was not much consolation to ISH to be able to say 'We told you so'.

Faced with poor trading results, troubles in the plant, (including the nuisance of fume pollution from the arc furnace which had attracted the unfavourable notice of the local health authority), rising costs, an unacceptable level of productivity, the chagrin caused by the verdict of the Restrictive Practices Commission, the acute and unresolved problems of adjustment to Common Market entry and the bleak short-term commercial prospects, it is not surprising that the company appeared to be in some disarray as 1972 drew to a close.

The report and accounts for 1971–72 had been presented to the annual general meeting in late November. They showed a net profit of £175,000 representing a decline of nearly £200,000, which was described by the chairman as 'a dangerously low return'. Turnover had been maintained by record exports of £1.7 million but these sales were, for the most part, at poor prices. No hope was given of better results in the following year, which would be the third and last of the difficult years of transition, and in fact a further drop in net profit was foreseen. Among the depressing facts noted in the report, one of great sadness was the loss of the talented deputy general manager, E. A. Coleman, who was killed in the London air disaster in June 1972.

Possibly it was an effort to enliven the gloom or possibly it was a sense of history that caused the chairman to include in his review at that meeting a salute to the company's silver jubilee. 'The omens were not propitious', he stated, 'when in 1947 the government decided to give the 'kiss of life' to the apparently dead steel works on Haulbowline and there were those who prophesied (and some who had an interest in hoping for) an early funeral.' The company had survived, however, and the review quoted some cumulative figures for the twenty-five years to illustrate what had been achieved from 'such modest and hazardous beginnings', and spoke of the contribution the small Irish industry had made to the economic welfare of the community. These figures seem to merit inclusion in a history of ISH and are as follows:

1947–1972	£
Capital expenditure – by the state (shares)	5,785,000
Capital expenditure – by the company	3,126,000
Total sales	60,811,000
Export sales	9,084,000
Wages and salaries	12,397,000
Expenditure on Irish materials and services	13,570,000
Net profits before taxation	2,397,000
Depreciation provided	3,364,000
Other reserves accumulated	1,852,000

Such a collection of figures does not prove any point in economic analysis. They might have been better or they might have been worse. There are no comparable standards against which they can be set, because the circumstances of the steel industry were unique in the Irish context. However, as an illustration of the results from a quarter of a century of endeavour, the figures present an impressive substance when related to a company which began life in 1947 with a paid-up capital of £3, a borrowing facility limited to £300,000 and a very uncertain future.

CHAPTER XI

A Reluctant European

The story of Irish Steel, if it is to be complete, necessarily includes the emergence of the European Community as a factor in the affairs of the company at Haulbowline. It was a gradual process, beginning as a possibility subject to debate and eventually becoming a hard fact of life.

In 1951, mainly due to French initiative, the European Coal and Steel Community (ECSC) was born. It comprised France, Germany, Italy and the Benelux countries. The event did not seem to matter to Ireland where coal mining and steel making were relatively minor industrial activities. The purpose of the Treaty of Paris, however, was designedly a first step in the objective (strongest in Germany) of Western European integration at all levels. The immediate aim was to promote the orderly expansion and modernisation of coal and steel production and to provide better conditions of living and employment for the workers in these industries.

It was a different matter for Ireland when, after protracted negotiations, the same six countries in 1957 formed the European Economic Community. The objects were stated in the governing Treaty of Rome. The task of the Community was to set up a common market, to harmonise progressively the economic policies and activities of member states, to promote continuous and balanced expansion and closer relations between member states. The Treaty made provision for the accession of new members provided they too were willing to promote the elimination of economic barriers and differences and to accept the Community institutions. In 1967 these institutions were unified, comprising one European Parliament, one Court of Justice, one Council of Ministers and one Commission or executive. Members were thus required to surrender some degree of national sovereignty to the Community institutions on which all the member states were to be represented.

There was a strong lobby in Ireland inside and outside government for joining Europe both on general policy grounds and also to break away from economic dependence on the British market. It was not an easy choice because the partial surrender of sovereignty was disliked in some quarters and also because considerable difficulties, mainly trading but also political (in the case of Northern Ireland) would arise if Ireland joined the Community and Britain did not. Britain had abstained from both the Treaties of Paris and Rome and had, in fact, established a separate Free Trade Association with the Scandinavian countries and Austria, Switzerland and Portugal. Ireland kept aloof from that association but later negotiated a separate agreement for progressive free trade with the United Kingdom (1966).

These were, of course, big issues by which the Irish steel industry was bound to be affected. The ISH directors, after 1956, were not unaware of the threat which the Treaty provisions offered to a steel company dependent on tariffs and government support for its existence. It is strange now to recall that when Sean Lemass conversed (briefly as was his wont) with the ISH chairman at a social function some twenty years ago, he asked, 'What do you think of the Common Market?' and received the reply, 'I think it will be an uncommon nuisance.' Admittedly the flippant rejoinder represented only the narrow views and fears of ISH. Lemass understood this when he added, 'There will probably be plenty of time for you to prepare for it.' It was indeed to be a long process and Lemass, so staunch an advocate of Ireland's role in Europe, did not himself live to see the day of accession to the Community.

In April 1957, the Department of Industry and Commerce officially requested ISH views on the question. The board's reply was that the industry in its existing situation would be unable to meet free competition under the terms of the Treaty of Paris. They suggested either deferment for a number of years or else some special arrangements under an agreement for association which would still permit Irish steel production to be protected in its home market. When Lemass visited Haulbowline on 2 October 1957 to discuss development possibilities, one of his arguments was the need to prepare for free trade in steel. Hogan and O'Shea were still sceptical as regards Irish involvement in the Community while Britain continued to stand apart. They really thought the EEC problem would go away. Development was necessary on other grounds.

During the years 1961 to 1963 Britain changed its attitude towards the Community and commenced negotiations for entry, but these were, for the time being, frustrated by a political decision taken by General de Gaulle. The French attitude also shut the door on other applicants for membership — Ireland, Norway and Denmark. The Irish application for full membership of the EEC was not formally lodged until 15 January 1963, when the government hoped that Ireland might escape the General's embargo on *les Anglo-Saxons.* This was not to be. Only in 1970 did the political difficulties inside the Community disappear so that the way was clear for formal negotiations leading to accession.

It was fairly obvious that the official view at the time in Kildare Street was one of anxiety about ISH affairs in the context of Europe. In January 1963 the board considered once again, this time under notice that the government had formally applied for full membership of the Community, a departmental request for information as to 'whether the steel industry could survive in common market conditions'. The board's reaction to this stark question was to reiterate their difficulties and the need for an adequate transitional period of protection for the steel industry. No worse time could be chosen for assessing the future. The company was overwhelmed by the unsatisfactory situation of the development project and by its current trading losses.

The steel company could not complain of lack of time for adjustment. After 1963 the 'period of grace', if such it may be termed, lasted for nearly a decade. During the interval ISH continued its efforts to establish a viable export trade in response to the government's Second Programme for Economic Expansion. Haulbowline Bridge was built, which amongst other benefits facilitated ISH efforts to adapt itself to Community requrements. One of the main reasons for ISH involvement in the abortive Project Ivor was the desire to arrange for the sale of the bulk of the company's products to an assured external buyer. When that scheme failed the interim development was undertaken as a means of giving the industry a better chance under Community conditions, a result which was frustrated by the extraordinary distortion and imbalance of the world steel trade from 1973 onwards. Efforts were repeatedly made by ISH during the years in question to establish some form of satisfactory association with an external partner which would strengthen the Irish steel industry for the hazards of trading on equal terms against the powerful producers within the European Community. After 1966 the company had experience of the effect of dismantlement of tariffs under the Anglo-Irish Free Trade Area Agreement. This proved less painful than had been feared, partly because of the troubles of the British steel industry after nationalisation. All in all ISH were still in a distinctly vulnerable position when in 1970 the white flag was raised and the race into Europe had begun.

At that time, notwithstanding Farren's exploratory visit to Brussels in May 1962, the Community hierarchy probably knew little enough about the prospective new steel member. A few years earlier, following an educational address on Community objectives and administration given in Dublin by M. Coppée, then head of the European Coal and Steel Community, the chairman of ISH had been introduced to the speaker. M. Coppée expressed surprise, 'I did not know Ireland had a steel industry.' Somewhat chastened, Hogan explained the situation of the insignificant industry and mentioned his company's anxiety about adaptation to ECSC requirements. He, in turn, was surprised when M. Coppée told him not to worry unduly; the rules were not being assiduously observed in every respect by the existing members. But he added grimly, 'We have had to fine some of them.'

Negotiations and Entry

In November 1970 the chairman informed Industry and Commerce of the company's views on the concessions for the steel industry which were to be sought in the pending negotiations with the Community. These were to be the longest possible period of transition, a gradual dismantling of tariffs, anti-dumping provisions and retention of the power to control exports of domestic scrap. It was also suggested that a representative of ISH should be included in the Irish negotiating team as the company was the only steel producer and had special problems. While this could not be conceded without embarrassment in relation to other interested parties, agricultural and industrial, equally concerned with the outcome, it was arranged that ISH be kept fully informed of the progress of the negotiations and that their advice would always be welcomed by the negotiating team. Close contact was thus maintained with the Department during the progress of the talks in Brussels.

Apart from the negotiations at government level which dealt with major issues, opportunities arose for ISH, as a producer, to have independent discussions with Community officials for the purpose of clarifying the small print of pricing systems and other controls. In December 1970 Coleman (Farren being ill) visited Brussels to participate in a discussion in which representatives of the steel producers in Britain, Denmark and Norway also took part. This meeting had been called to enable the applicant producers to discuss any difficulties they foresaw in meeting the requirements explained by the Community officials. Coleman's opening words when called on to speak: 'Ireland has no problems,' astounded his listeners and were a cause of some merriment at the lunch-time break.

This was the first of several visits to Brussels by Coleman. In May 1972 the general manager conducted the discussion with the Steel Community and reported to the board about outstanding points of difficulty in complying with ECSC regulations. There was a problem for the Community in applying its pricing regulations to Britain and Ireland, islands relying on sea-borne trade in which freights were various and variable whereas, within the Six, internal movement of steel had been solely across land frontiers, with well established rail and river or canal freight rates being applied in the formulation of approved selling prices.

In October 1972 the chairman and general manager went to London for talks with senior officials of both the BSC and BISPA. The interchange was valuable in yielding information as to British intentions in adapting their marketing and pricing practices to Common Market conditions. Friendly contacts were established which would prove useful to ISH in future years. Also during 1972 the original four applicants for membership had been reduced to three, Norway having dropped out because of the somewhat unexpected decision of the people when their government submitted the proposal to a referendum. In the case of Ireland, a referendum was also essential. Ratification of the Treaty of Accession involved legislation which could not be enacted without prior amendment of the Irish Constitution under the provisions of which any proposed amendment has to be referred to the people. The referendum was held on 10 May 1972 and resulted in a large majority in favour of accession. It was urged that much more would be gained from increased agricultural exports at better prices than might conceivably be lost in employment terms from exposing a relatively weak industrial base to the hazards of free trade competition. That was the economic argument which, added to political preference, produced the huge vote for joining the Community.

In September 1972 officials of the Steel Community visited Haulbowline for the first time. The discussions were helpful, although it did appear that some of ISH practices could be out of line with Treaty of Paris principles; notably there were those which might infringe the anti-discriminatory provisions governing marketing and pricing (e.g. concessions to manufacturers and small provincial buyers) and those which might create a dominant situation in a particular sector (e.g. as regards scrap purchasing). In November there were further talks in Brussels for the purpose of ensuring that the pricing system which ISH intended to adopt with Haulbowline as the basing point for Irish steel prices, would be acceptable to Brussels. In all these matters of detail the company was concerned not alone with its own interests but also with those of its home customers who were called upon to adjust their practices to the new system of trading.

At the annual general meeting on 22 November 1972, the chairman, referring to the pending accession on 1 January 1973 when ISH would become a steel enterprise governed by the Treaty of Paris, reported, 'In the negotiations for accession our government secured concessions for the steel industry, including easements of the rate and timing of reductions in protective tariffs and retention of scrap export control, which will operate during a transitional period of five years. These dispensations will be of considerable benefit to us in meeting the problems of adjustments.'

It was not all plain sailing however. Towards the end of December it was necessary to inform Industry and Commerce that ISH was not ready for the accession date. Discussions were still going on with the officials in Brussels and with the National Association of Steel Merchants. One of the problems was that ISH, which had always sold its steel at the works, could not establish comparable prices with other Community producers unless it included delivery prices in its published price lists. To achieve this, following negotiations which lasted until May 1973, CIE was appointed as carrier from Haulbowline to all destinations in the Republic at agreed rates. This arrangement was accepted as satisfactory by the National Association. If, however, a customer elected to pick up his order at the works with his own transport, he was required to pay a 'collection extra' of £2 per ton, this extra charge being in line with the practice followed in similar circumstances by other Community steel works. A few customers found the 'collection extra' vexatious, but only one firm persisted in refusal to pay the charge, which obliged the company to take legal proceedings to enforce its rights.

A more serious matter arose from the apparent intention of the government to maintain control over the level of steel prices charged by ISH. During the second half of 1972 and the early months of 1973, Western Europe prices for steel and steel-making materials began to move upwards sharply. When ISH wished to increase its prices to meet this situation, application still had to be made to the Prices Commission although on the accession date the High Authority of the ECSC had become the price controller for Irish steel under the Treaty provisions. ISH protested about this situation at first without success but eventually, when there had been some nudging from Brussels, the Minister for Industry and Commerce formally notified the company that its products were exempt from national price control in accordance with the requirements of the Treaty of Paris to which Ireland had subscribed. Therefore, from August 1973 onwards ISH could freely adjust its selling prices in line with market conditions, subject to publication of its new prices in accordance with Community regulations. It was an important gain in commercial flexibility, the profitable effect of which was seen in the company's improved trading results for 1973–74.

Shortly after accession ISH received an invitation to join the Steel Club (*Club des Sidérurgistes*). This body was a long-existing association of steel producers within the ECSC somewhat, but not admittedly, resembling the old steel cartel of the thirties. The Club's purpose was to agree joint policies and to make known the producers' views to the Community Commissioner in charge of steel. Each of the existing six countries in the ECSC possessed a national federation of steel producers embracing many companies. It was these national bodies which sent delegates to the Club. The three new members were in a different situation. Britain's steel industry was mainly, in terms of volume, controlled by the nationalised BSC. The small privately owned steel works had banded together in the British Iron and Steel Producers Association (BISPA). Both BSC and BISPA were members of the Club, just as they were also separately represented at committee and other meetings within the Community. In both Ireland and Denmark there existed only one steel company, so it was the company and not a national association which became a member of the Steel Club.

The board accepted membership of the Club. It was inadvisable to do otherwise although it was apparent that a small producer such as Ireland could exercise little influence and that the policies and interests of the big powers would dominate proceedings and conclusions. Except for crisis occasions the Steel Club met quarterly at both director and president levels. The general manager of ISH on some occasions attended the directors' meetings, but more often it was P. Mortimer who did so. The chairman went to the presidents' meetings whenever possible during 1973 and 1974 with either Farren or Mortimer also in attendance. These occasions involved visits to Brussels or Amsterdam or Dusseldorf, invariably sandwiching long meetings between uncomfortably hurried journeys.

Apart from the value of making personal contacts with the leaders of the other Community steel industries, the business of the Club did from time to time involve questions of concern to ISH such as the efforts to reach arrangements with the authorities in Japan for voluntary limitation of their steel imports into Europe, the measures taken in the USA to curtail Community steel imports and the Community intentions to negotiate a trade agreement with Spain involving special concessions to the Spanish steel industry which included a quota of Community scrap. The latter was a sore point with ISH who were refused facilities to purchase continental scrap during the transitional period of scrap export control in Britain and Ireland. At the presidents' committee in March 1974 the ISH chairman pointed out that it was inequitable for a non-member country to be given a scrap quota when it was being denied to a member of the Community, particularly at a time of scrap shortage; in this contention he was supported by the representatives of France, Italy and Denmark. The proposed agreement came to nothing in face of the steel producers' opposition to the terms affecting their interests until eventually an application by Spain (1977) for full membership changed the situation. In the meantime ISH continued to have difficulties in importing scrap except for inferior grades licensed for export from Britain. The embargo on scrap exports to Ireland was, however, lifted after 1975.

At the ISH annual general meeting in December 1973 a reference was made to administrative burdens and expenses imposed on the steel company by membership of the Community. 'I do not complain of this,' said the chairman 'because we must play our part as a steel producer within the Community.' The additional expenditure included levies on ISH production payable to ECSC funds under the terms of the Treaty. These payments amounted to a total of about £78,000 during the first five years of membership, a period of low production at Haulbowline.

Payment of the levies is balanced by the company's access to special funds provided by ECSC for improvements in the steel industry which meet the criteria laid down. One of these schemes provides cheap loans towards part of the cost of houses for workers in the European steel industry. Under this scheme the workers in ISH qualified for loan assistance and the Commission were willing to make funds available. So far

128 new houses and thirty-seven reconstructed dwellings for occupation by Irish steel workers have qualified for loans under this scheme.

At the instance of the Department of Industry and Commerce, ISH applied late in 1973 for a Community grant towards its expenses on researching a scheme to preserve and enlarge employment. The conditions which the ECSC attached to such grants were, however, found unacceptable by the ISH directors, mainly because they would or could involve a breach of confidentiality between ISH and other interests with whom plans for development at Haulbowline were being considered at that time. It was probably as a result of the disclosure of these negotiations in the ISH memorandum of application for the research grant that two senior officials from Brussels visited Haulbowline in May 1974. They were careful in discussions with the chairman and Farren not to appear in any inquisitorial role but only to offer helpful observations. They emphasised that the Commission would be concerned to ensure that any project for joint venture between ISH and an external partner did not involve a concentration or domination of the market contrary to the provisions of the Treaty of Paris. The Commission feared that ISH might be submerged by some big producer — indeed BSC was hinted at — and it was indicated that in their own interests ISH should consider partnership with some small or medium-sized operator, whenever such association became ISH policy. There was no Commission objection to ISH having a partner from outside the Community. It was possible to infer from the visitors' guarded remarks that it would be Commission policy to cooperate in maintaining the Irish steel industry and that any plans for expansion would be considered in that light.

On 12–13 September 1974 the Bureau of the EEC Consultative Committee on coal and steel held a meeting in Ireland for the first time. The Committee and the Bureau both comprise representatives of producers, consumers and trade unions from the nine Community countries. Their important function in the coal and steel community (Art. 7 of the Treaty of Paris) is 'to assist the High Authority', now the Commission of the EEC. As the only steel producer in Ireland, ISH was given one seat on the Consultative Committee. This was filled first by Paddy Mortimer until his retirement in 1978 and then by Gerard Farren.

ISH entertained the Bureau delegates to a luncheon. Speaking at the function the chairman took the opportunity to suggest that, in considering the problems of the steel industry as a whole, the Commission should give heed to the importance of preserving and strengthening steel production in every member country. In Ireland, only one such producer existed. It should be helped even though its production was a minor element in the overall Community steel economy.

Crisis in the Steel Community

At the end of 1974 the Irish steel industry was beginning to suffer acutely from cost inflation and from the persistent depression in the home market for its products. The steel producers in other Community countries were also in trouble in varying degrees, least so in Germany. Reporting on a meeting of the President's Committee in Brussels on 26 November the ISH chairman noted the general spirit of pessimism about short-term prospects and the lack of agreement as to the necessary corrective measures and the order of priorities. There was a move to advise the Commission that, in order to conserve raw materials and economise on energy requirements, expansion plans based on electric arc (EA) furnaces or on direct reduction methods should be viewed with disfavour. ISH had spoken against this proposal (which was talked out) because it affected the Irish development plans which contemplated EA furnaces and the use of natural gas, recently discovered off the southern Irish coast, in a direct reduction process.

At the ISH annual general meeting in December 1974, when a trading profit (before depreciation charges) of £1.32 million was reported, the report referred to the company's pricing policy under Community conditions. The object had been to minimise the domestic inflationary effects of high steel prices. The margin sought on turnover had not exceeded 10 per cent but even this degree of profit was no longer attainable in the face of cost inflation, including the effects of greatly increased oil and electricity prices and of market recession. If this trend continued, Irish prices could become uncompetitive within the Community.

The situation of the Community steel industry worsened during 1975. The representatives of ISH continued to participate in discussions as to remedial measures both at the Consultative Committee of the ECSC and at the Steel Club. They also talked directly to Community officials on the special problems of the Irish industry. In his report to the ISH annual general meeting on 18 December 1975 Kevin McCourt described the nature and extent of the serious problems facing the whole Community steel industry. He related the dramatic fall in production and sales during 1975, the distortion of selling prices, and how the steel plants in the nine ECSC countries were in aggregate operating at 60 per cent of their capacity. The chairman said that ISH had to contend at home and abroad with offers of foreign steel at low prices which in almost every case were quoted in contravention of Community rules or, where non-EEC countries made the offer, were really dumping prices. McCourt took the EEC Commission to task for its failure to protect the Community's steel industry. Remarking that some of the Treaty protocols, having been negotiated at a time when steel demand exceeded supply, may have been designed to safeguard the consumer rather than the producer, he went on, 'There is, however, a provision under Article 58 of the Treaty of Paris enabling a manifest crisis to be declared in circumstances such as exist at present. Measures vested in the Commission when a crisis has been declared include power to control production, to fix minimum prices within the Community, to impose tariffs or import quotas against non-Community steel. Despite strong pressure by some European steel producers including this company, the Commission in Brussels has so far failed to take decisive steps to cope with this admittedly complex situation.'

The condition of the Community steel producers went from bad to worse during 1976. The hope for recovery of demand, such as had happened after the temporary depression of 1970, was not realised. The producers, as Kevin McCourt had hinted in his December 1975 statement, were not unanimous in their pressure on the Community for direct action, while, for its part, the Commission was hesitant about taking any formal steps to safeguard the steel industry which would run counter to the principles of free trading on which the Treaties of Paris and Rome were founded. The obligations of member States under GATT were also at issue. In August 1976, the Commission issued an analysis of the causes of the steel crisis and of the steps which might be taken voluntarily by the producers to correct the situation. In September Gerard Farren attended a meeting between the Steel Club and Commissioner Simonet concerning proposals for a temporary reduction in outputs by all Community producers. Following this meeting, the Commission, while abstaining from a mandatory 'Decision', issued guide-lines as to the extent to which production should be reduced, as a percentage of outputs in the previous year. These reductions were to apply universally to all producers, but after representations by ISH as to their low output in the previous year caused by domestic circumstances, the Commission agreed that the company should be exempt from the reduction applied under the 'Simonet plan' and should instead accept a guide-line keeping output of crude steel during the first quarter of 1977 to 10,000 tons per month.

The 'Guidance' restraints on production were an imperfect attempt to bring supply into better balance with demand. It was an attempt that did not solve the problems of cut-throat pricing within and without the Community, often in contravention of the Commission regulations as producers struggled to maintain sales from the reduced outputs. In early 1977 the Commission issued a Decision imposing minimum prices for reinforcing bars and it also indicated Guidance prices for other steel products. Later on in November 1977, the Commission sought to impose, by formal notice to national associations of steel merchants, an obligation on merchants to refrain from buying steel at prices below the minimum fixed by the Commission directive and guide-lines. Finally on the last day of 1977, after debate at the Council of Ministers, new measures to protect the steel industry were published. The range of obligatory minimum prices were extended. A levy on low-priced imports from outside the Community was introduced. The legal obligations on steel merchants and stockists were intensified. As a help to producers the Guidance prices imposed a year earlier were increased.

The foregoing outline, necessarily imprecise, of the Community efforts to solve the problem, excludes measures taken by the Irish government within its own limited powers to protect the native steel industry. Towards the end of 1976 offers in Ireland of non-EEC steel at low prices were becoming a serious threat. In November of that year, ISH asked the statutory authority (*Coimisiún Dumpála*) to impose anti-dumping duties against steel products from Spain, South Africa and Japan. Following some delay, while evidence of dumping within the statutory provisions was being obtained, the Minister for Industry and Commerce on the advice of the Comisiún Dumpála imposed special duties operative for six months against imports of reinforcing bars and angles from Spain (January 1977) and against imports of bars, rods and angles from South Africa (March 1977). No such action was taken against Japanese imports, apparently because of voluntary steps by the authorities in that country to curtail the level of their steel exports to Community countries. The evidence that anti-dumping duties would if necessary be imposed helped to dissuade Irish importers from placing orders with countries outside the EEC.

During October 1976 the chairman of ISH had broken new ground by attending the annual meeting of the International Iron and Steel Institute, held in Osaka on that occasion. ISH had become an associate member of that body some years earlier so as to have access to the papers and statistics it publishes on technical and economic aspects of the world steel industry and to the reports of proceedings at the annual meetings attended by representatives of all the leading steel producers. The Institute has no executive or advisory functions, but has an important role as a meeting ground for interchange of ideas and for forecasts as to the future trends in the industry and in the nature and extent of new investment. At Osaka there may have been some optimistic voices as to recovery in demand and the ending of the crisis but as Kevin McCourt said at the annual general meeting in December 1976 this unfortunately was a mistaken view.

At the end of 1976 the Steel Club was replaced by Eurofer as the representative body of Community steel producers. The essential purposes of the new body in relationship with the Community remained unchanged but it had been restructured to meet the evolving situation of the steel crisis. All the ten members of the old Steel Club signed an agreement at Amsterdam on 9 December 1976 which established Eurofer as a legal entity under the laws of Luxembourg, which was to be its headquarters with Memorandum and Articles of Association and a board of directors. The purpose was to form a stronger administrative bureau and a better system of sub-committee consideration of individual problems in the industry, thus enabling Eurofer to speak with a more influential and authoritative voice in making representations to the Commission. At the top is the *Conseil d'Administration* or board of directors, which differs from the Presidents' Committee of the Steel Club. Under the earlier system there was equality of representation whereas now there are four directors on the *Conseil* from Germany, three each from Britain and France, two from each of the lesser national groups tapering down to one each from Denmark and Ireland. However, each country still has only one vote. The ISH general manager, Farren, had signed the agreement setting up Eurofer and he became a first director.

It is too soon in 1978 to evaluate the experience of the small Irish industry as a member of the coal and steel community. Only briefly after accession were trading conditions normal. That period coincided with the 'growing pains' of compliance with the Treaty of Paris pricing and other requirements. Then the economic crisis supervened, world-wide in its scope but most seriously affecting the security of the Community steel industry. Thus the greater part of the five-year transitional period which in different circumstances would, as intended, have given Irish industry a chance of gaining market strength and technical improvement was occupied in a struggle for survival. The misfortunes of ISH were not such as could have been effectively alleviated by retiring within a protectionist wall – that is, on the hypothesis that Ireland had not joined the EEC and was free to adopt such measures. Being a member of the steel Community during the years of crisis at least has given the Irish steel industry the benefit of association with a much larger group in facing a common peril and in adopting practical measures to maintain some degree of viability while the crisis lasts. The testing time will come when improved trading conditions for steel producers have been stabilised over a number of years and when it can be seen how far membership of and support from the Community will ensure continuance of a small national steel industry on the perimeter of the European economic system. Doubtless it is for ISH itself to establish by its own efforts the technological and marketing adjustments, supported by progessive outputs, worker efficiency and management skills, which will give the industry a secure future within the European Economic Community. That having been said, however, it is also true that much will depend on the policies and measures adopted by the Community institutions in giving effect to the principles and obligations stated in the Treaties of Association. Among the latter are those which are designed to prevent undue concentration of economic power and domination of markets.

CHAPTER XII

The Years before the Storm

The tyranny of dates suggested that ISH had reached a jubilee occasion by the middle of 1972. The directors, on being so reminded, found themselves ill-disposed towards any measures of celebration. The recent trading results and the company's anxieties in facing the Common Market did not encourage euphoria. On the other hand it was six years since Haulbowline Bridge had been opened with some panache, and nothing had yet been done to mark by official ceremony the completion of the interim development at a cost of some £3.5 million which the company had been able to meet from its own cash and credit resources. The Taoiseach, Jack Lynch, accepted the chairman's suggestion that he should, on 26 January 1973, inspect the new installations and be the company's guest of honour at a celebratory luncheon.

Jack Lynch saw in action the new electric arc furnace and its ancillary equipment, the extended soaking pits and the improvements effected in the rolling mills. On his way through the plant he conversed with managers and workers. As a gesture of encouragement from the Head of Government, the visit was a marked success as was the subsequent luncheon. The company had assembled a multitude of guests to do honour to the Taoiseach and, in a lower key, to mark a jubilee occasion. These included the Lord Mayor of Cork, deputies and senators, chairmen of various public bodies, as well as old friends of ISH in the steel trade both at home and abroad and in other businesses and professions, many of whom had in one way or another been concerned in the activities of the steel industry since its birth. The good health of the Taoiseach was drunk with acclamation and he, in turn, proposed a toast to the prosperity of the industry, speaking about its achievements and its importance to the national economy.

Some time afterwards a cynic observed that it was a mistake for a Taoiseach to attend an ISH luncheon of public character. There had only been two such occasions. Shortly after the first in 1966 Sean Lemass had resigned his office. Within a month of the second, Jack Lynch's party had been defeated at the polls and a new Taoiseach, Liam Cosgrave, took office as head of a Fine Gael/Labour Coalition. The change brought Justin Keating, a Labour deputy in Dublin, to the Department of Industry and Commerce. Justin Keating was popularly regarded as one of the intellectual stars of the new administration. Qualified in agricultural and veterinary science, a university lecturer in Dublin and a television personality, he had made his mark in politics as a front-bench opposition spokesman. When the ISH chairman made his *ad liminem* visit to the new Minister, Justin Keating asked many penetrating questions and promised a visit to the industry at an early date. On 8 June 1973 he came to Haulbowline, inspected all the operations and discussed future projects at length with the directors and senior management. He encouraged the company to adopt dynamic measures of expansion. The money would be found if a development scheme of viable quality could be established.

During 1972 and 1973 ISH affairs were dominated by the problems of the Common Market described in the preceding chapter. On the home front trading conditions were exceedingly dull and no sign of improvement emerged until the first half of 1973. Export possibilities were restricted and, in the main, export margins were small. At the annual general meeting on 5 December, the chairman reported a small net profit of £59,000 in 1972–73 on a record turnover of nearly £9 million of which £1.7 million had been gained from exports. This unsatisfactory profit had been dictated by adherence under price control to low selling prices during a period when, in conditions of high demand coupled with a steep rise in raw material and labour costs, other countries had increased their prices. The Irish domestic customers had substantial benefit from ISH prices which were so much below the international level. The price freeze at Haulbowline was, however, not the sole reason for the poor profit. Production targets had not been reached owing to a series of mishaps in the new plant, to the inferior quality of imported scrap and to a low level of productivity from a large labour force. None the less, the chairman offered hope that in the following year, production, sales and profits would reach new records for the company.

The most important element in that report was possibly the reference to future plans for development, still only at the study stage. It was necessary to adopt the most modern technical methods and equipment, suitable for a steel works of moderate size but able to sell competively at home and abroad. Expansion of electric furnace capacity and substitution of continuous casting of billets for the more costly operation of ingot casting, soaking pits and heavy rolling mills were prime desiderata, but consideration would also be given to direct reduction of imported ore to a form suitable for charging to electric furnaces, thus relieving dependence on scrap metal. It was clear that such a large capital investment could not be undertaken from the company's cash flow and reserves and in due course proposals would be submitted to the government. In a reference to the advantages of Haulbowline and Cork Harbour as the location for an expanded steel industry, mention was made of the discovery of natural gas off the South Coast and its possibilities if made available for steel-making.

In December 1973 when the foregoing remarks were publicly made about ISH development targets, the company had for some time been engrossed in negotiations of a more or less promising nature. Over the years it had been the policy of the directors, endorsed by successive Ministers, to welcome proposals for external participation in the industry provided that the intending partner fulfilled some tests of eligibility. These were summarised in November 1973 as follows:

1. Willingness to risk substantial capital;
2. Provision of skills and expertise not readily available to ISH;
3. Control of/or access to raw materials;
4. Control of/or access to markets.

It was difficult to foresee the emergence of any candidate who would fulfil all these qualifications. On balance the first and last were the most important.

In mid-1973 the Industrial Development Authority introduced to ISH a German industrialist who was anxious to secure full or partial control of steel supplies from Ireland and was prepared to make an investment. Erich Frei operated a substantial merchanting business and also processed and fabricated steel sections for his trade customers, He had no experience of steel production but had assembled a few technical advisers from the German steel industry to assist him in his plans for joint venture with ISH in expanding operations at Haulbowline. The scheme was labelled Project Cormorant (suggesting a shy, low-flying bird) but this tendencious title of convenience did not prevent the ideas put forward by Frei from being discussed in depth and hopefully at a number of conferences both in Ireland and Germany. There was a visit by an ISH delegation including an IDA representative in October 1973 to inspect the Korf miniworks at Hamburg which used natural gas in a direct reduction plant as well as operating continuous casting. They also saw the merchanting and processing plants controlled by Frei at Bremen and Brunswick, clearly efficient and progressive operations.

Project Cormorant was never formulated in sufficient detail to enable its cost and probable viability to be accurately assessed. The scheme had attractions if, as represented by the German interests, the partner would buy all the products of a new merchant mill and guarantee adequate supplies of German scrap. On the other hand, as discussions proceeded, the extent of the intended Frei investment became doubtful, while the difficulties involved in the suggested separation of the old ISH works including steel-making, in which Frei would have no share, from a new Cormorant Works (and Company) in which he would invest 50 per cent proved considerable. But for the interest of the IDA, it is doubtful if negotiations would have been as protracted as they were, involving the employment of engineering consultants to investigate the cost of site development on reclaimed land and of new deep water wharfage for the increased tonnages envisaged. It was not until June 1974 that negotiations were ended by mutual agreement. By that time more attractive suggestions for possible partnership had arisen. Moreover, the directors had concluded that only firm and definite proposals for external investment should be allowed to delay expansion of the industry on the lines recommended by their own technical staff.

In the late summer of 1973 a prominent Irish industrialist approached the ISH chairman with confidential information that he proposed to establish a steel-making and rolling plant elsewhere in Ireland. A modern mini plant of about 150,000 tons capacity was envisaged at the outset. Mr P (for such we will call him) invited ISH to invest in the company he had formed to undertake this new industry. The immediate reaction was to point out that with ISH intending to expand its output and range, there would be no room for two steelworks in Ireland. There would be cut-throat competition for domestic scrap supplies and for the Irish market in merchant steel. The ISH chairman suggested that if Mr P wished to diversify his other activities into steel, he might be better advised to invest money in development at Haulbowline. Mr P's scheme, which was necessarily secret, was noted in the minutes as Project Puffin and there were further discussions when Mr P proposed that he and ISH might combine to establish a direct reduction plant at or near Haulbowline based on natural gas. The ISH attitude was that with Project Cormorant still a very live issue, realistic consideration of the various elements of Project Puffin, all of them still tenuous, was impracticable. This by no means implied lack of interest in Mr P's ideas, to which he returned from time to time in the next few years. His financial strength was undoubted, as was his proven success in other commercial and industrial fields.

The prospective availability of natural gas in Ireland was also the mainspring in the next approach to ISH. The managing director of Sheerness Steel Company in Kent, the most recent and highly successful mini-works in Britain and one outside the nationalised BSC system, wrote to the ISH chairman in November 1973 advancing his company's interest in development at Haulbowline. Discussions followed with Sheerness and with Co-Steel Inc. of Canada which had promoted and held the controlling interest in the British company. Both were interested in direct reduction based on natural gas and were prepared to assist ISH as partners in a major development at Haulbowline if technical and marketing studies demonstrated viability within the European Community system. The ISH board welcomed this possibility, all the more so when it was found that the chairman of Co-Steel was Gerry Heffernan who, as related in chapter eight, had come to Haulbowline in 1962 as one of the two Canadian engineers commissioned by Industry and Commerce to furnish a report on the First Development. Heffernan now came several times to ISH for discussions. It was agreed that Sheerness and ISH should jointly finance a comprehensive feasibility study by Canadian consultants. This was begun in November 1974. As to natural gas, no scheme involving that facility could be considered unless the Irish government altered its declared intention to make the reserves of gas so far discovered available only for electricity and fertiliser production.

On 21 October Justin Keating received the ISH and Co-Steel chairmen, who were accompanied by Farren and Schueppert, managing director of Sheerness. The Minister was encouraging and said that if the negotiations produced a project for major economic expansion of the Irish steel industry and if use of natural gas was a decisive element in such a scheme, he would propose to the government that the required quantity of gas should be diverted from use in electricity generation. This assurance satisfied the promoters that they could proceed towards finality as soon as the consultants' report had been completed. In December ISH issued to Co-Steel a formal letter of intent to proceed to joint investment as soon as the arrangements contemplated, scale of development, financial costs and partnership sharing, had been mutually agreed. That was the situation at the end of the period covered by this chapter. An important element in the proposed development was that a direct reduction plant of minimum economic size would have outputs in excess of ISH requirements and that Sheerness would be responsible for using or marketing the surplus. As the situation evolved after 1974, the ISH directors decided to exclude direct reduction from their plans for essential short-term development.

Although much smaller in scale than the other projected developments, there was another proposal to which ISH gave consideration during 1973 and 1974. The future of the galvanising and corrugating plant at Haulbowline caused concern, partly on account of its being now semi-obsolete and partly because of the inevitable removal under Community regulations of the preference enjoyed by the Irish product under the government scheme of grants for agricultural buildings.

Hence there was an advantage to be gained when the BSC expressed willingness to consider a joint venture with ISH in the erection of a modern galvanising and corrugating plant with sufficient capacity to serve market requirements in the whole of Ireland. A preliminary market survey was financed jointly by both parties. At a meeting with BSC officials in London on 8 February 1974, the ISH chairman and Farren reached understanding as to the objectives, assuming viability could be established. The new plant would be erected on ISH property on the mainland, and would be owned by a separate company to be financed and controlled in equal partnership by the Irish and British interests. There was much promise but little fulfilment in these plans. Viability became elusive in relation to costs and profitability. By the end of 1974, although still alive, the galvanising project no longer made great appeal.

In February 1973 the general manager had submitted to his board a comprehensive review of the position of the industry. It was a depressing document, as regards plant performance, productivity, trading experience and financial outlook, while at the same time, the Common Market problems still called for solution. The board urged the need for adequate programmes of preventive maintenance and stated its readiness to employ additional technical staff for that purpose. The financial problems were eased by a grant of £800,000 from the IDA towards the company's expenditure on the interim development project. While trading conditions did not improve in time to affect the results for the year ended 30 June 1973, the financial position was no longer a worry during 1974. The company's situation as a heavy borrower from the banking system changed to one of holder of interest-bearing deposits. Unfortunately, this comfortable position lasted for too short a time and altered dramatically in 1975 for the reasons stated in the next chapter.

As to plant performance, the board minutes of June 1973 record a statement by John Rafferty who was in charge of the works that a solution of the production problems 'would take time and patience'. He never ceased to urge that, regardless of the engrossing proposals for external participation, ISH must in any event secure its future by expending capital on, as a minimum, the introduction of continuous casting and the replacement of open-hearth steel production by additional electric furnace capacity. This was the general view of the company's technical staff, who were told by the board to prepare their own scheme parallel with the consideration of other projects for development involving external interests and probably greater investment. The object was to have in being an alternative plan which could, with consultancy advice, be rapidly brought to the project stage if the partnership ideas came to nought.

During the decade which opened in 1970, questions arose affecting all commercial companies conducted under state auspices. The report of the *Public Services Organisation Review Group 1966–69* (known as the *Devlin Report*), made radical proposals to the government for the reform of the public services. These proposals drew into the net of the executive tasks of Ministers not only public bodies financed by parliamentary grants but also the commercial state-sponsored companies which, operating in the competitive market economy, had not hitherto been regarded as part of the public services.

The report stated that 'the creation of these bodies was motivated by the desire to give them commercial freedom and to have them operate in a commercial environment' (C. 13.5.2), and that the 'goals' of such a body having been defined by the government, 'their attainment is a matter for the board' (C.13.5.6.). Such pronouncements were in keeping with what might be described as the Lemass concept of the autonomous responsibilities of boards of directors of state-sponsored commercial companies. That concept has been exemplified in the instances recounted earlier in this book of ministerial relations with the directors of ISH. The report, however, went on to recommend new practices which seemed to erode the traditional autonomy of such boards. 'Commercial state-sponsored bodies should be linked to the Aireachts (the policy-making units in the sponsoring departments) which should actively engage in the definition and review of goals, the appraisal of results and the control of capital expenditure. Each functional assistant secretary should have responsibility for this and should, therefore, maintain continuous liaison with the chief executive of the body concerned' (C.13.5.10).

These proposals for continuous monitoring implied a

degree of departmental control of the commercial companies which had not previously been found necessary or desirable. The objective of 'commercial freedom' seemed to be pushed into second place. The roles both of the board and of the chief executive responsible to it were much affected. The board appeared to lose some of its power and influence. The chief executive seemed to be required to wear a reversible cloak – responsible to the board and answerable to the assistant secretary.

In July 1972, a second report was published, that of the *Review Body on Higher Remuneration in the Public Sector.* That body was guided by the same chairman as the earlier *Review of Organisation,* so that the two reports were referred to as *Devlin 1* and *Devlin 2.* The second report recommended the adoption of pay levels for the higher posts in the public sector as a whole, including local authorities, ministerial and parliamentary offices and the judiciary. In the case of state-sponsored commercial companies, specific limits of pay were proposed for the chief executive of each company according to the report's evalution of the comparable importance of the various bodies. These limits were to be mandatory on the appointing boards and were to include not only actual pay but also the value of any benefits, such as a company car, which might be received by the chief executive. The proposals therefore meant that the board could select its chief executive but not decide how much he was worth to the company in terms of salary. The most suitable person (possibly the only one suitable) could not be appointed if he were unwilling to accept the limit of pay prescribed by *Devlin 2.*

These recommendations in the Devlin reports caused concern to the board of ISH. The executive director, in sending comments to the departments, strongly urged the view that inclusion of a commercial company such as Irish Steel among 'the public services' and thus within the domain of the executive functions of government would be a mistake. He stressed the deterrent effect of the Devlin recommendations on the promotion in state enterprises of the spirit of initiative and risk-taking which were essential to commercial success. The disadvantage which such an arrangement would impose on these enterprises in the comparison (which the Devlin report suggested should be the test of their success) of their performance with that of similar enterprises in the private sector was obvious. So also was the danger that the restraints on remuneration of chief executives in state companies would be a disincentive in the recruitment of staff generally and condemn these companies to an artificial system of salaries out of line with the market forces which determined how talent and experience should be rewarded. Among the functions of company directors were those of hiring, firing, and remunerating their staff including above all the chief executive on whom they placed the primary responsibility for the success of the undertaking. That function was now to be curtailed without, it was claimed, any compelling reasons.

The government, however, accepted without qualification the recommendations in both Devlin reports regarding commercial bodies and took steps to implement them as a measure of reform in the organisation and objectives of the public services taken as a whole. Time will tell whether such arrangements will be appplied with the degree of flexibility required by the commercial companies.

During 1973 and 1974, following government adoption of the Devlin prescriptions, the ISH board fruitlessly sought approval, firstly from George Colley and later from Justin Keating, for somewhat better renumeration for their chief executive than the arbitrary limits allowed. In the case of ISH the chief executive for this purpose was the general manager and not the executive director who, by terms of appointment and contract, was not a whole-time officer of the company. At the end of the period covered by this chapter, the directors were still seeking to find some way out of the strait-jacket imposed by the Devlin formula. (A note on the administrative structure of ISH appears at the end of this chapter.)

Another matter of government policy which involved the board during these years was the intention of the Minister for Labour to include ISH among the state companies which, as a measure of social and industrial reform, would be required to include a number of workers in the industry among the appointed directors. The ISH board did not question the general principle involved in the proposed legislation, which although different in degree and structural base, harmonised with similar arrangements in some other Community countries. They pointed out, however, that such a proposal was ill-timed in the case of the steel industry and might frustrate possible partnership with external interests. The directors also pointed out that ISH were already in advance of most other companies in the public sector through the existence of their Management/Labour Works Committee. Through the good offices of the Minister for Industry and Commerce, the ISH case was accepted and the company was excluded from the scheduled bodies in the Act passed to give worker participation at board level a trial spin. This was not the end of the matter. The Minister for Labour was to reopen his proposals affecting the steel industry in a few years' time.

Shortly after the annual general meeting of December 1973, the executive director gave notice to the board that he intended to resign that office on completion of his existing contract in the following March. He was aware that the government wished, as opportunity offered, to replace elderly directors in the public sector by younger men. However, the Minister for Industry and Commerce asked the ISH chairman to remain in office until the next annual general meeting in December 1974, by which time it was hoped a suitable successor to the chairmanship would have been found. The chairman agreed to do this as a final service to the industry. It could scarcely be claimed that his retirement was premature. He had been a member of the board for twenty-seven years. The sprint which had been briskly undertaken by Williams, O'Shea and himself in 1947, had turned out to be a mile event and then became a marathon for the sole surviving runner of the three; it was high time to seek respite and pass on the baton.

The report and accounts for the year ended 30 June 1974 were announced at the annual general meeting on 17 December. The heads of the Departments of both Finance and Industry and Commerce, C. H. Murray and P. Ó Slatarra, attended on behalf of their respective Ministers, as a valedictory tribute to the outgoing chairman and as a welcome to his successor-designate, Kevin McCourt, whose formal appointment took place during the meeting. Hogan's promise of record results, made a year earlier, was fulfilled by the company's report. On a turnover of £14 million, a trading surplus of £1.3 million was gained. After making increased provision for depreciation that surplus became a net profit of £846,000. This profit was sufficient to provide for a maiden dividend if that had been feasible. The Articles precluded payment of dividends but in any event it was expedient to retain the money for future capital development. The balance sheet was strong, showing reserves of £2.65 million, accumulated depreciation of £3.1 million and a liquid surplus of over £4.3 million.

In an unscripted statement to the general meeting, the chairman gave warning of the unsatisfactory trends since the end of June 1974. In the first five months of the new accounting year a modest profit had been earned, although galvanised sheet sales had fallen by £2 million compared with the same period in the previous year, causing the company to suspend production and disemploy workers in that sector of the industry. The chairman foresaw reduced turnover and profits over the whole year but his forecasts proved wide of the mark. As it turned out in 1975, it was no longer a question of enduring reduced profits but of facing substantial losses.

The business of the meeting included the re-appointment of G. van der Puil as director of the company. He had been appointed in April 1974 by the Minister thus bringing the board to the maximum number allowed by the Articles. G. van der Puil, managing director of Verolme Cork Dockyard, with much experience of industrial challenges similar to those facing Irish Steel, was well equipped to assist the steel company and, as he amply proved, willing to do so.

A few days before the annual meeting an unusual event occurred. The links between the steel industry and the town of Cobh have already been stressed. On 12 December 1974 the Urban District Council convened a special meeting which the Cork County Manager, Michael Conlon, attended for the purpose of receiving from the ISH chairman the title deeds of a building site in Cobh which the company owned and had decided to donate to the local authority for use in erecting chalets to be occupied by senior citizens of the town. There was a secondary Council purpose in calling such a special meeting. They wished to give a Civic Address of Welcome to the ISH chairman, which was intended and received as an exceptional honour. Pat O'Mahony, the chairman of the

Urban Council, presented the illuminated scroll of Address, which as well as tributes and good wishes to the recipient conveyed warm recognition of the company's contribution to the prosperity of Cobh and, in particular, of the gift of property 'which will be an enduring monument on Cobh's terraced hillside to the magnanimity of Irish Steel Holdings'.

Note on the administrative structure of ISH (see p. 190)

Aspects of the managerial and executive organisation of ISH have been incidentally mentioned in the course of this history. This note fills some gaps as regards both facts and people, bringing the record reasonably up to date.

The main managerial burden has necessarily fallen on the occupant of the post of general manager, of whom there have only been two. Christopher Fitzpatrick was succeeded late in 1955 by Gerard Farren whose achievements have been mentioned frequently in the story of the intervening years. His ability, loyalty, and strong character have for nearly a quarter of a century been tirelessly devoted to the interests and policies of the steel industry. To other industries he also made a contribution during that time, having been a part-time director of both the Shannon Free Airport Company and the Verolme Cork Dockyard Company.

When E. A. Coleman occupied the post of deputy general manager, for not much longer than a year prior to his tragic death in June 1972, there was some relief for the usually over-burdened general manager. Subsequent to that the board adopted a new organisation of senior management. They created five posts of assistant general manager with powers of decision-making under appropriate delegation from the general manager. The intention was to leave more time for the general manager for his essential tasks of planning future strategy and of formulating policy proposals for consideration by the board of directors. Like all such blueprints for organisational perfection this one did not work in the manner or to the degree intended. It was related to reasonably normal conditions, which never seem to have prevailed at Haulbowline.

The managerial organisation so formed in 1973 brought to these new posts five able men who had had considerable experience of the industry since their first recruitment. John Rafferty, already works manager, who had grown up in the melting shop, became AGM (works); Jack Adair, the chief accountant for many years, became AGM (finance); Harry Kerr, an engineer by calling and with ISH since 1958, became AGM (personnel); Charlie Vaughan, recruited in 1962 and with long service as secretary of the company, became AGM (procurement and planning) and Paddy Lynch, Coleman's former assistant, became AGM (sales). This team was strongly supported in the technical field by George Radley (mills manager), Reg Warbrook (planning controller) and George Reynolds (chief engineer), all experienced professional engineers. In addition to the foregoing, the senior management included, from 1973, P. F. Mortimer as 'manager, ECSC department'. Coming to ISH from a higher post in Industry and Commerce, he was well fitted to assist the company in the problems of adjustment to Community membership.

Adair's protracted illness made it necessary in 1977 to recruit a financial controller, this being L.S. Coughlan, formerly of Clover Meats Ltd. John Rafferty's untimely death late in 1978 brought Michael Walley to be his successor as AGM (works). In the reorganisation caused by Gerard Farren becoming 'director, ECSC affairs and economic planning' in the Dublin office and by the subsequent appointment of Kevin McCourt as executive director, Vaughan and Coughlan were appointed deputy general managers at Haulbowline. The post of general manager was vacant, presumably a temporary situation.

As is true of other heavy industries, steel works are served predominantly by men. However, lest the absence of any woman from the names so far mentioned may create suspicion of discrimination, let it be stated that women do find employment at Haulbowline. Not, indeed, on the factory floor, although women are so employed in steelworks elsewhere, mainly in quality control. The clerical and secretarial staff at ISH comprises a high proportion of young women, some of them the indispensable confidential secretaries to senior executives. Moreover, there is among the staff one woman of high rank, Miss Ita O'Reilly of Cobh, whose manifest abilities are at present (1979) applied to the responsible post of financial accountant.

During a period of thirteen years, from 1961 to 1974, the administrative organisation in the steel industry was necessarily affected by the existence of an executive director who was also the chairman of the board. He was virtually a managing director because the board had given him such extensive delegated powers. The advantages which the executive director's functions and presence were intended to give to company performance were twofold. He was, whenever necessary, available for consultation by management and he was also in a position to make immediate decisions on matters which otherwise would have awaited the briefing and meetings of the full board. This arrangement was revived in 1978 when Kevin McCourt became executive director.

A survey of the ISH administrative scene would not be complete without some reference to the links between the company and the government departments concerned for its welfare. There were two of these: Industry and Commerce, and Finance. The departmental responsibility was either general, in the context of overall policy in relation to commercial bodies in the public sector, or particular as regards matters (mainly financial) which by the Articles of Association or by legislation were subject to Ministerial determination or consent. Responsibility for the day-to-day affairs of the company belonged to the directors.

In the early days of ISH, matters requiring communication or consultation with the departments were readily settled in discussion between civil service directors and their departmental colleagues or sometimes, on important matters, by interviews with Ministers. After the time of Williams, MacCarthy and Connor as ISH directors and after Hogan had retired from Finance in 1961, communication between the company and departments, when not handled by the executive director, involved to a greater degree the general manager and other senior officials of the company. This change took place at a time when, due to a variety of causes, the affairs of the steel industry engaged departmental attention to an increasing degree. When the responsibilities of the ISH management require assessment, they necessarily include the fairly continuous two-way communication with Industry and Commerce, which no doubt will be intensified under the Devlin Report concepts.

In Industry and Commerce, the main voice, on major issues at least, has usually been that of the assistant secretary dealing with ISH affairs. The occupants of that post have been men of ability with a sympathetic understanding of the problems of the steel industry. It is not, one hopes, invidious to mention in that context Andrew Kennan and John Donovan, who both had, of course, many other claims to be regarded as distinguished public servants. This author has much reason to be grateful to these gentlemen for their support and counsel whenever he sought it during his days with Irish Steel.

CHAPTER XIII

Bright Prospects Succeed Crisis

At the beginning of 1975 the steel works on Haulbowline stood virtually unchanged since completion of the interim development. To a casual observer, moreover, the layout and essential features of the working areas did not materially differ from the scene a decade earlier, except for the additions of the bridge and the warehousing facilities on Paddy's Block. One might also have observed the disappearance since 1968 of the two lofty red-brick chimney stacks which were so conspicuous in early pictures of the industry. Having become unsafe they were taken down and replaced by metal flues which, however efficient and economical in maintenance, certainly lent no enchantment to the view.

The office quarters still occupied the old Admiralty Storehouses, nos. 10 and 11, which faced east towards the works and the basin. Their thick walls and stone floors and stairs, the latter with wrought iron handrails, betrayed the origin and purposes of these massive houses when built in the early years of the nineteenth century. In such a habitat it is unremarkable that ISH management occasionally tended to be victims of a siege mentality, but at least they escaped accusations of executive luxury.

It was from these offices that the visitor of 1975 vintage, bedecked with steel hat and white coat and accompanied by an escort, emerged to begin a quick tour of the works. Turning south and proceeding between the boundary of the naval property and the storage premises and waste-heat boiler house erected by ISH on the filled-in east camber, he would arrive at the passenger pier and Spencer Jetty, a walk of some 500 paces. Reversing direction, he would then enter the main works building which housed under contiguous roofs all the production processes of the industry. Steel-framed and roofed and clad with galvanised sheeting, this complex is probably the largest such factory area in the Republic, of sufficient strength to support the many overhead travelling cranes, including one with the ability to lift over one hundred tons as required by the casting bay. The factory is 300 yards in length and covers about seven acres. The product flow is from north, where the raw materials for steel-making are assembled and prepared for the furnaces, to south where the finished steel emerges from the rolling mills and is made ready for dispatch to customers' requirements and where the long road bridge links the island to the mainland.

The northern half of the great factory was occupied by the melting shop (a 60-ton oil-fired open-hearth furnace and a 30-ton electric arc furnace and its circuit breaker), the casting bay and the soaking pits, while alongside the melting shop was the scrap-yard and its wharf. The southern half of the building housed the rolling mills and ancillary equipment, the despatch department, the pickling, galvanising and corrugating equipment, the laboratory and quality control services together with the underground electricity sub-stations. If the visitor had surveyed all these operations, had climbed and descended the many steep and narrow steps, had stood close to the furnace heat or to red hot ingots and billets or bars travelling through the stands of the rolling mills, and had sniffed the sulphuric acid in the pickling bath, then he would not be unhappy to leave the factory building at the southern end close to the bridgehead and there breathe cooler fresher air, while resting his smarting eyes on the vistas of Cork harbour. His journey was not yet completed. He must cross the main works thoroughfare which links the bridge with Spencer Jetty, and visit the engineering workshops and the works canteen before returning to his starting point after perhaps forty-five minutes if he had been conducted at a rapid pace and had not proved more inquisitive than a casual visitor ought to be. Had he been more curious or energetic he could have observed the results of reclamation at the eastern tip where scrap reserves are stored and Slag Systems Ltd. conduct their metal recovery work, or he could have crossed the bridge (bowing en route to the historic remains on Rocky Island), inspected the warehousing facilities and surveyed the acres on the mainland which formed the company's land bank for possible industrial use in the future.

No steel works is a place of beauty nor even of enjoyment for the visitor intent on a tour of inspection. Haulbowline could not claim any exemption from that judgment because of an American tribute that it is the most beautifully situated steel plant in the wide world – a remark offered as a solace in the context of less complimentary comments! Individual visitors or groups of industrialists or students, always welcome when they give due notice, were carefully shepherded by management so as to avoid any danger while at the same time they were given a full picture of the industry. Steelmaking involves many spectacular and arduous operations in conditions of heat and noise, and it has none of the trim and tidy smoothness of say an electricity plant or a textile factory. Haulbowline did not qualify for high marks in appearance or even cleanliness, yet few visitors went away unimpressed by the sheer weight and dynamism of the relatively small Irish works and by the collective skills and endurance required from the teams who operated the furnaces and the mills.

The somewhat nostalgic pen picture of the 'old works', for such they will soon be termed when the extensive changes begun in 1978 have been completed, represents the industrial area and equipment which those in charge of the steel industry had at their disposal at the end of 1974. They also had an experienced management and a work force, including administration, of some 1,100 people. Neither the physical nor the human assets could be fully employed in the circumstances which developed at a surprising rate in the early months of 1975, as markets receded and cash flow dwindled. It was no passing phase, nor one which derived its momentum solely from the Irish economic recession and the troubles of the construction industry. Equally harmful was the world-wide contraction in production and trade which afflicted the international steel industry, especially in Western Europe.

The survey (in chapter eleven) of the first faltering steps of the Irish steel industry inside the Common Market necessarily covered many of the events and problems of the period between the ISH annual meeting in December 1974 and the end of 1977. This final instalment of the story of over forty years of endeavour and survival at Haulbowline cannot be readily understood without adverting to the circumstances described in that earlier chapter.

It was thus a time of great difficulty and some danger that faced the board of directors. Some of these men were new entrants into the story of Irish steel. Kevin McCourt, chairman of the company since December 1974, brought to the steel industry great experience and a long and varied record of success in the management and direction of important undertakings in Ireland and for a while in the Netherlands. Formerly secretary of the Confederation of Irish Industries, later a first director of the IDA and more recently a member of the board of Foir Teoranta, Kevin McCourt added extensive knowledge of other industrial problems to his own experience as director of many financial and commercial companies, not excluding his distinguished role for several years as director-general of the National Broadcasting Service. In 1968 he became group managing director of Irish Distillers, which was his main commercial activity at the time he agreed to add the leadership of ISH to his other heavy responsibilities.

In April 1975, when Ben Devlin had retired after nearly a decade of valuable service, Justin Keating appointed Harry Bambrick to the board of ISH. He was managing director of Whessoe (Ireland) Ltd. and also a member of the Labour Court panel. Late in the same year the Articles of Association were altered so as to increase the number of directors from five to seven, thus enabling the Minister to appoint two more experienced industrialists. John Byrne B.E.—whose years may have reduced a little the average age of the ISH board—had recently relinquished the chief executive post in CIE in order to engage in other professional and commercial pursuits. Dr A. C. Cusack had for long been a leading figure in Irish industry as managing director of the Clondalkin Paper Mills and their allied companies. Throughout a critical period in ISH fortunes, therefore, the direction of the industry was guided by a most experienced board—McCourt, Frame, Dillon, van der Puil, Bambrick, Byrne and Cusack—whose combined talents and successes covered many and varied fields. The departure of Devlin had severed the last link with the so-called 'civil service dimension' on the board, a feature of the company's early history. In December 1978 when Dr

Cusack and Professor Dillon had retired, the vacancies were filled by Paul Conlon and Sam Clark, both men of experience in other Irish industrial enterprises.

During the years of crisis the major problems faced by these directors and by all those engaged in the steel industry at Haulbowline may conveniently be grouped under three headings viz. trade and finance, industrial relations and plans for modernisation and expansion. These were concurrent problems the last of which was the most important, being the key to survival and prosperity.

Finance and Trade (1975–78)

From July 1974 onwards, sales of galvanised sheets had virtually ceased. After November the market for other ISH products was seriously affected. The company's results for the first half of the financial year 1974–75 did not fully reflect the adverse trend. A net profit of £300,000 was reported and the liquidity position remained satisfactory. The recession in demand, however, worsened rapidly beyond even the most pessimistic expectations.

During the course of 1975 there was a world-wide collapse in the steel market such as had not been experienced for forty years. As the year wore on, the board were faced with two serious problems: first, the inevitable cut-back in production with consequent redundancies; and secondly, the necessity to increase short-term borrowings as cash flow progressively fell below outgoings. For a while the deposit funds at high interest rates (at one point as much as £1.4 million) which had been accumulated during the previous two years gave protection. By August 1975, however, all these deposits had been used for existing needs. The overdraft facilities available from the Bank of Ireland were supplemented by borrowings from other institutions, including Algemene Bank of the Netherlands and the First National City Bank, which discounted the steel company's short-dated bills of exchange. In May 1975 the board had directed that steel production and manning be reduced. As a result the 70-ton open-hearth furnace was taken out of commission, *sine die,* thus anticipating the demise of that installation under the company's programme of modernisation. With market conditions as they were, the electric arc furnace was more than ample for production needs.

The results of the financial year ended 30 June 1975, submitted to the annual general meeting in December, showed that turnover, compared with the previous year, had fallen by 23 per cent to £10.8 million. The volume of sales was down 39 per cent. The serious decline in home sales (45 per cent) had been offset by an increase in exports amounting to £2.6 million but the prices obtained abroad had reflected the depressed market conditions. So it was that the net loss for the year amounted to £685,000 compared with a net profit of £846,000 in 1973–74. The loss was covered by transfers from reserves. The chairman gave no hope of immediate recovery. It had been decided to suspend production from the electric arc furnace for six months because ingot stocks in the works were sufficient for rolling requirements during that period. Kevin McCourt observed gloomily, 'It cannot be another year of loss . . . significantly greater than that reported today.'

The decision to suspend steel-making as a temporary measure affected the purchase of scrap. Formerly, even at the worst times, the company had continued to accept all domestic scrap offered by suppliers. This policy sometimes, as during the First Development, had involved buying and holding immense stocks of scrap rather than weaken even temporarily the system of export prohibition. At the end of 1975 the company was in no position to pursue that policy. Consequently the board decided that purchase of scrap should be restricted to current requirements and reserves not exceeding 12,000 tons; that export licences would be recommended for dealers holding scrap not accepted by ISH; and that because of the export relief, a reduction in the price of scrap supplied to Haulbowline should be sought. The last condition was a faint hope although it may have delayed the price increase which the scrap merchants were seeking. An increase of £5 per ton is recorded in June 1967.

The unhappy outcome of the hearings before the Restrictive Practices Commission in 1972 has already been mentioned. Under duress ISH had introduced the scrap purchasing system recommended by the Commission and endorsed by the Minister under his statutory powers. As ISH had foreseen, the system had proved a failure. In 1976 the board formally asked the Minister to relieve ISH of the obligation to continue it. They also requested him to prevent the illegal export of scrap. At this point the board were, no doubt, anxious to ensure orderly control of domestic scrap supplies not only for current needs, but also in preparation for the development of the industry which would depend on greatly increased consumption of scrap as the basic raw material.

As the economic storm increased in intensity during 1976 and showed no signs of blowing itself out, the sufferings of ISH became almost intolerable. It is a matter for admiration that, in such circumstances, the general manager and his staff maintained cool heads and high resolve. No opportunity of expanding production and sales was lost. It was quite a triumph that significant and continuous export sales went to Germany in face of keen competition from both inside and outside the Community. The still depressed home market was stimulated by extending the product range to include flat sections in the rolling programmes and by offering a new type of deformed reinforcing bar, 'Tor Bars', for the manufacture of which ISH obtained licensee rights in the Republic from the British patent holders.

The results of the financial year 1975–76 showed a net loss of £1.268 million on a reduced turnover of £9.3 million after all normal provisions had been made, but this was increased to over £2 million by the addition of 'extraordinary items', in effect a surgical operation to prepare for development, whereby certain plant, buildings and spare parts were the subject of special depreciation. The large loss was mainly covered by transfers from the Taxation Reserve and from the General Reserve (then exhausted) but there still remained £287,000 as a deficit to be carried forward. As to the future, Kevin McCourt said, 'Because of the uncertainties as to world economics, as to the demand for steel and steel prices as well as to the effectiveness of the EEC measures, it is difficult to make predictions . . . I hope, however, that because of an anticipated increase in demand, coupled with measures aimed at improving productivity within the company, we will achieve better results than those reported for 1975–76. . . . This will also depend upon maintenance of harmony in industrial relations throughout the company, because without this the many difficulties in production and trading cannot possibly be contained even to the limited extent within our control.' As it turned out the last point in the chairman's conditional hope for improvement was a vital one. As he was speaking in December 1976, the threat of strike action by a key section of the work force was already a dark and menacing cloud.

A long-standing dispute between ISH and its one-time supplier of pig iron at fixed prices under forward purchasing arrangements was eventually settled. The London firm concerned had failed in 1972–73 to meet its contractual obligations due, as it claimed, to *force majeure*. Having had to accept pig iron from another source at higher prices, ISH suffered financial loss. Legal proceedings were eventually begun in London to enforce the ISH claim for damages due to non-fulfilment of contract. The case was settled out of court by a substantial payment to ISH plus costs. In the 1976–77 accounts the amount of the compensation thus received from the supplier is noted as £177,383.

In spite of the continued difficulties both trading and financial, 1977 might have shown an improvement in ISH fortunes had the works not been closed for so long. There were signs of an upswing in the Irish economy as a whole, with a welcome though modest revival in the construction industry from which the steel company was unable to benefit until the last quarter of the year. Unfortunately the financial problems of the Haulbowline company, already serious, became intense due to a domestic cause, the loss of revenue while strike action prevented production. In these circumstances temporary borrowings on overdraft and bill reached a new peak. In addition to the institutions mentioned earlier, the Citibank NA and the Ulster Bank provided short-term accommodation.

The depressing experience of ISH was related on 21 December 1977 when the report and accounts were submitted to the annual general meeting by the chairman. Although turnover had increased by nearly £2 million, the trading loss was about £320,000 higher than in the previous year. After allowance had been made for further special depreciation of assets and for other extraordinary items, the net loss shown was slightly in excess of £2 million, all of which had to be carried forward to the next account.

The report referred to the strike of craft union workers lasting nearly six months (March–September 1977). Actual trading losses directly arising from the strike were about £700,000. As to the immediate future, Kevin McCourt could hold out little hope for much improvement in the financial situation, although a revival in home trade orders, particularly for galvanised sheet, gave encouragement. 'It is impossible for me', he said, 'to project other than a further loss for the year 1977–78.'

That indeed was the outcome revealed at the annual general meeting in December 1978. None the less, some improvement was recorded. Turnover had increased by over £2 million during the year; because of price inflation this was a gain of only 11 per cent in volume of sales. The trading loss was reduced by £348,000 but when heavy provision had been made for special depreciation of existing plant, the net loss on the accounts was shown at £1.668 million. Accumulated losses carried forward after four years of storm and stress had thus reached a total not far short of £4 million.

These unfavourable financial results were, however, balanced by the better news which the chairman gave to the meeting: the company's development plans had received the full backing of the government and the measures to implement them were already in train. At long last, light appeared at the end of the tunnel. The meeting was also heartened by the chairman's announcement that at the Minister's request he had accepted appointment as executive director in addition to the chairmanship, thus further committing his great experience and energy to the success of the development project.

Industrial Relations (1975–77)

In the story of Irish Steel during the early and mid seventies, the problem of over-manning in the industry has been mentioned more than once. That situation, aggravated by wage inflation, existed even when the plant was fully and continuously operated. The financial effects became acute when production fell away in face of the severe market recession that showed no signs of early recovery. Moreover, an additional cause of concern arose from the fact that the company's development plans, although involving greatly increased outputs, depended for their viability on the employment of a permanent labour force far smaller than the numbers employed at the end of 1974.

What had to be faced, therefore, was not merely the temporary lay-off of men whose employment was interrupted, as had happened occasionally in the past because orders were low or raw material deliveries had been held up, or plant was under repair, or had been taken out of commission while new equipment was being installed. These were the normal hazards of industrial employment. In such circumstances the long established trade union practice of 'last in, first out' was applied. Now what had to be considered was the permanent disemployment not only of recent recruits but also of workers who had longer service with the company.

In March 1975 the board directed that measures should be taken in consultation with the affected Unions to reduce employment by a scheme of voluntary redundancies applicable to those with the requisite length of service. This entailed offering severance terms better than the redundancy payments provided by statute in cases of permanent disemployment in businesses which had been curtailed or abandoned. The system of voluntary redundancies thus involved ISH in heavy payments in order to secure both its objective and the agreement of the Unions, who, naturally, pressed for the best possible terms in the interests of their members.

The number of voluntary acceptances of redundancy was one hundred and seventy-eight. The statutory payments in such cases amounted to £102,000. The additional payments by the company were £196,000. After recovery of the state contribution of over £68,000 towards the statutory payments, the total burden falling on ISH was about £230,000. The amount of each individual redundancy payment varied greatly depending on the length of service and on pay at the date of disemployment.

As a result of the voluntary redundancies, the laying off of temporary or casual workers and the non-filling of vacancies caused by death or retirement where this was practicable, the ISH labour force was drastically reduced. Loss of good employment by so many people was a grievous consequence of the measures forced upon the company during its struggle to survive a period of unprecedented difficulty and to prepare for better times in the future. The average number employed during 1974 had been 1,124. (This figure included both men absent through illness and casual employees.) Two years later the average number had fallen to just over seven hundred. If the painful contraction of the labour force was unique in the company's history, so also was the official strike of the craft workers to secure increases outside the limits allowed by the National Wage Agreement. As Kevin McCourt said in his address to the ISH general meeting in December 1977, 'It should not be assumed that this strike reflected the normal climate of industrial relations within the industry. In fact, over the thirty years of our existence, the general pattern has been one of reasonable peace and harmony; problems have been settled by negotiation rather than industrial action.'

In ISH the craft workers, all skilled in one or other of the trades in which they had served their apprenticeship, were employed in the essential work of maintaining and repairing plant and equipment. They were, relative to the whole work force, few in number but they were members of vigilant trade unions and, at Haulbowline, were represented by active shop stewards, some of whom it would not be unfair to call militant in their outlook.

Since 1971, the limits governing pay adjustments had been governed by successive National Wage Agreements entered into between representatives of Unions and employers. These arrangements were an effort to avert the hazards and inequalities of 'free collective bargaining' during a time of rising inflation when the public interest and national policy required control of incomes and prices. Such National Agreements contained provisions allowing special increases above the normal limits where 'anomalies' in wage structures either generally or in a particular business were shown to exist or where management and workers entered into genuine productivity agreements involving additional pay for additional output. The Agreements provided machinery for settlement of disputes, including reference to the Labour Court, and for avoidance so far as possible of strike action. The foregoing represents in very rough and ready terms the environment in which the dispute between the company and its craft workers was conducted. Although it must be added that on the company side there was the additional background of its own financial situation and its obligation as a state organisation to adhere to the terms of the National Wage Agreement in force at that time.

The craft workers' claim which, when unsatisfied, was reinforced by strike action, arose from a comprehensive regrading of the jobs of ISH general workers pursuant to a Labour Court Arbitration Award. The award entailed increases in the scale pay for the various grades. In February 1976 the craft unions stated that the re-grading had disturbed relativity. They sought for their members a 20 per cent established differential above the top scale applicable to general workers. This claim was investigated by the Labour Court which recommended payment to the Craft Workers of a fixed bonus of £2.19 per week with retrospective effect to 24 February 1974, the same date from which the general workers' re-grading became effective. That recommendation was accepted by the company but rejected by the Unions, who wanted much more and prepared for strike action. In an effort to avert a strike, the company proposed that negotiations be commenced for a productivity agreement but this was flatly rejected by the shop stewards.

The craft workers withdrew their labour on 29 March 1977 and placed pickets. This was an official strike. The Unions which supported it were: the Amalgamated Union of Engineering Workers; the Electrical Trade Union; the National Engineering and Electrical Trade Union; the Union of Construction Allied Trades and Technicians; the Amalgamated Society of Boilermakers, Shipwrights, Blacksmiths and Structural Workers and the Ancient Guild of Incorporated Brick and Stonelayers Trade Union. About ninety men went on strike, the bulk of them being either electricians or fitters. Their action went on long enough to paralyse the industry and cause disemployment of most of the other workers including some office personnel. Many of these and their families suffered considerable hardship from loss of wages.

The company sought guidance from the Adjudication Committee of the Employer/Labour Conference (the parent of the National Agreement), which suggested that work should be resumed on the basis that the Labour Court award be implemented and that the company should table specific productivity proposals. This suggestion was rejected by the Unions whereupon the Adjudication Committee declared that the strike action constituted a continuing breach of the National Wage Agreement (22 April 1977).

Notwithstanding the inflexible – apparently 'win at all costs' – attitude of the Unions and the shop stewards, the company persisted in its search for a solution within the National Agreement. Abortive conciliation conferences, under the aegis of the Industrial Relations Officer of the Labour Court, were held in May and again in June. In June, too, the Lord Mayor of Cork and the Chairman of Cobh Urban Council held separate meetings with company officials and with the shop stewards but without result. The Minister for Labour offered to refer the dispute to the Labour Court for a special investigation under the Industrial Relations Act on condition that normal working was first resumed. The Unions did not reply to that offer. In July, meetings were held at the Irish Congress of Trade Unions Headquarters in Dublin. The Congress were naturally alarmed at the duration of the strike and its implications. Again the outcome was fruitless.

Between 26 August and 6 September, at meetings in Cork under the Labour Court Conciliation procedures, the deadlock was eventually broken. As is usual in such cases, where a strike is protracted, neither side could claim a victory. Too much had been lost by the long closure for this to be possible. The company maintained its position that any settlement must be within the scope allowed by the National Agreement, but it had to pay more than the Labour Court had recommended a year earlier and to agree to a productivity deal that was certain to provoke claims from the general workers. The craft unions for their part did not secure the 20 per cent differential for which they had undertaken the strike but they gained more than the company had been prepared to concede at an earlier stage.

It has been calculated that the annual addition to the company's wage bill for the craft workers was £59,000. In the context of the total labour costs in the industry it does not seem a heavy price to pay for industrial peace. Nor does it seem reasonable that a strike of such long duration, and with such ends to achieve, should have been embarked upon, with all its lamentable effects on an already stricken company and inevitable hardship to many individuals. There were other consequences, however, for the company that were not included in the bill of £59,000 annually. The settlement with the craft workers stimulated wage demands from the general body of ISH workers who in their turn argued that relativity had been disturbed. Such claims could be entertained only with the virtual certainty of reaction from the craft workers.

The details of that calamitous and unprecedented strike are an essential part of recent ISH history. They also present a case study in the difficult industrial relations which disturbed the Irish economy generally during the late 1970s.

Development

The genesis of the project for major development in association with Canadian and British interests has already been related. During 1975 matters did not progress in a satisfactory manner although the negotiations with Co-Steel and Sheerness Steel Company continued. On 18 March the chairman and the general manager visited Sheerness and discussed the problems with Gerry Heffernan, Schueppert and Sir Charles Wheeler. Doubts arose about a scheme based on direct reduction (DR) of ore with off-shore gas as the fuel. A month later it is recorded that the ISH board regarded involvement in direct reduction as 'premature' but were prepared to reconsider the question if Co-Steel/Sheerness insisted on DR as a condition of participation. This situation was reported by the chairman to the Minister for Industry and Commerce.

In the meantime the report of the Canadian consultants, Ferrco (who had an association with and had done much work for Gerry Heffernan's company), had been adopted in principle. This involved a decision that the expansion of the industry should proceed on the basis, at least in the short term, of using scrap as the steel-making raw material. These issues were again discussed at a meeting in London on 14 July but without conclusion as to how far the Sheerness company would participate financially. It is obvious that the Sheerness proposals on that score were unacceptable to

ISH, who, however, still retained confidence in the consultancy recommendations of Ferrco. In October 1975, the ISH directors decided to inform the Minister that they wished to go ahead with development without any external participation and would submit proposals based on the studies and recommendations of Ferrco as adapted in consultation with the company's technical staff. That decision represented an important change of course, dictated by the necessity both of early modernisation of plant if the industry were to survive in future competitive conditions, and also of proceeding at once to heavy capital expenditure even without a partner. 'Going it alone' on a commitment estimated at that stage to cost at least £30 million, including working capital, was a big step. The IDA, who were expected to provide a substantial grant and the Minister who had the ultimate responsibility of approval, may have looked askance at a project which had no external support. These hesitations may have determined some action taken during the early months of 1976 whereby the IDA and ISH jointly examined the feasibility of using natural gas at Haulbowline and discussions were renewed with Co-Steel on the question of participation. The ISH board did not, however, waver from their position that the scheme they had formally recommended to the Minister, reinforced by a meeting between the chairman and Justin Keating, was the right course to follow and was one which did not brook delay if both capital costs and predicted results were to conform to the forecasts.

In May 1976 van der Puil and Michael Walley, the ISH steel-making manager, visited the Chapparal Steel Works in Texas, originally designed by Ferrco for Co-Steel. In the following month the chairman had discussions with Gerry Heffernan both in Canada and Ireland, apparently making it plain that if there was to be participation it could only be in the scheme which ISH had recommended to its government as suitable for the needs of Haulbowline. In the result Co-Steel decided against investment in the ISH project.

ISH continued to be hopeful of an early government decision, in the meantime taking steps to prepare the ground. Ferrco were formally contracted as consultants for the execution of the project and authorised to incur expenditure up to £100,000 on preliminary design and layout work. The Investment Bank of Ireland were appointed financial advisers in relation to the project. Discussions were begun in Luxembourg with a view to obtaining assistance for the ISH project from the Community funds available for approved development in the coal and steel industries. It was not until February 1977 that the Minister for Industry and Commerce informed the chairman of ISH that the government had approved the project and that the company could proceed to implementation on the financial projections in its submission. Shortly afterwards Kevin McCourt at a press conference in Cork made public the objectives and the cost of the proposed development of the industry. The summary of his statement is included in a note to this chapter, as a convenient method of explaining the essential features of the most important development, in terms of cost and timing, in the company's history.

The hopes of making progress were frustrated by closure of the Haulbowline works because of strike action. While the strike continued, a further complication arose from a change of government in June 1977. It was apparent to the directors that in view of the long interval since their project had been put forward in January 1976, and taking into account the changed circumstances of the industry, none of them favourable, the project should be reassessed.

Kevin McCourt's statement on development when he addressed the ISH annual general meeting on 21 December 1977 reveals the position at that date:

> Because of the lapse of time from submission of its development proposals, the unexpected continuation and intensification of the depression in world steel markets and the need, of course, to assess the effects of the six-month strike on the company's finances and on customer relations, the board felt it wise to have its proposals re-examined by an appropriate independent organisation. The report of the consultants, A. D. Little Ltd. was received late in November.
>
> The consultants confirmed in large measure the company's projections upon which approval of the project was based. However, given the complexity of the project and the background against which the board had decided to commission this special examination, we deemed it appro-

priate that the Minister for Industry, Commerce and Energy should be given an opportunity of studying the report.

The report stresses that the earlier the contracts for development plant are placed, the better will be the company's competitive capacity to take advantage of the upsurge in demand for steel products in due course. We look forward to being in a position to proceed with the project in the very near future.

While the company were awaiting confirmation from the Minister (now Desmond O'Malley, who had replaced Justin Keating at the change of government) that the development proposals should proceed, an unfortunate circumstance arose. As required by the Treaty of Paris the company had at an early stage notified the EEC Commission of its development intentions, on which it was open to the Commission, if it so wished, to express an opinion. Such an expression of the Commission's views was sent by letter to ISH in the closing days of 1977. This was a confidential document but in some way it was 'leaked' in Brussels with the result that the Irish newspapers published 'scare' reports and comments to the effect that the EEC had vetoed the ISH proposals and that the Irish steel industry might have to close down. In fact the Commission had no power of veto, although their view of the ISH proposals was adverse, as regards both timing and the extent of the proposed increase in production, in the context of the Commission measures to reduce overall steel production within the Community while the depression in world steel markets continued. The Commission 'opinion' was not a final one, it suggested that ISH should consider development on a lesser scale and invited consultation on the whole question. The Irish government were, of course, also involved in the matter, as owners of ISH, and were in a position to make representations to the Commission, apart from discussions between the company, its advisers and officials of the Commission.

The EEC Commissioner, M. Davignon, whose responsibilities included the steel industry, visited Dublin towards the end of January 1978 for discussions with Irish Ministers on a variety of topics. His reported statements at a press conference about the ISH existing development proposals were discouraging although he made it plain that the decision rested with the Irish government. Undoubtedly the Commission's attitude was a severe setback to Kevin McCourt and his board. Not only did it necessarily involve delay while efforts were being made to reach agreement on the course to be followed by ISH with government approval, but it seriously prejudiced the prospects of securing financial assistance from the Community credit institutions. There was also the depressing affect on the morale of management and workers at Haulbowline of the misleading media publicity concerning the Commission's views and their possible effects. The turn in events was calculated to call in question the role of the Commission as seen from Haulbowline. Having survived a battering for three years and requiring the proposed development to secure its future, was the industry to be condemned because some 'Eurocrats' in Brussels took a view about its needs that was opposed to the views of the company and their experienced consultants? These were natural reactions by those who depended on the steel works for their livelihood, even if they were unfair to the Commission and its objectives. The chairman of ISH remained optimistic and instilled confidence; his resilience and combative spirit, proven so many times in his career, were more than ever necessary at this juncture in the fortunes of Irish Steel.

Commencing in February 1978 the company held a number of meetings with EEC officials and also took further advice from the consultants A. D. Little Ltd. Following upon these discussions, and under the auspices of the European Commission, ISH negotiated a scheme of production/marketing rationalisation with a French steel producer, *Société Métallurgique et Navale Dunkerque Normandie* (SMNDN). The two companies have agreed to market up to 30,000 tons per annum of each other's products as soon as the new Irish plant is operational in 1980. Rationalising production lines to suit the needs of each other's customers, both companies will be able to reduce production costs. As the projected annual output from Haulbowline gradually reaches by 1985 the target of 282,000 tons of saleable products there may be scope for extending the arrangements with SMNDN.

The conclusion of the negotiations with the French producers provided the European Commission with the opportunity to reconcile its views with the Irish proposals for expansion. The happy outcome was announced by the chairman at the ISH general meeting in December 1978. The scope of the development project so approved is in all essential respects the same as that described in the note at the end of this chapter. Cost had increased somewhat since Kevin McCourt spoke of it in March 1977, now involving fixed capital of £32 million and working capital of £9 million. There is firm expectation that ISH will obtain ECSC loans of at least £15 million and capital will also be provided by equity investment, IDA grant and banking facilities.

When the good will of the Commission had been secured and when the green light for progress had been so unequivocally shown by the government, ISH pressed forward. By November 1978, orders had been placed in Italy for a 90-ton electric arc furnace, in Canada for the continuous casting machine and with Liebherr of Killarney for several new cranes. As Kevin McCourt had stated it was his board's policy that as much as possible of the investment should be spent in Ireland. By January 1979 an Irish contractor (Ascon) was already engaged at Haulbowline in preliminary site works. While Ferrco of Canada remain the main consultants, Irish site and engineering consultants have been associated in the task of bringing the project to a timely and successful completion.

It can thus be seen that the year 1979 became one of hope and excitement. In spite of the difficulties which ISH continued to share with the Community steel industry, it was possible to look forward to bright horizons with virtually a rebirth of the steel works in Haulbowline. As the chairman had said in December 1978 when speaking of the development, 'it will provide security of employment and well-being for all who work there. While at the same time being internationally and domestically competitive, in addition it will keep pace with changing technology, repay its borrowings and service the State's investment of capital.'

With that statement of the objectives for which the momentous development project provided the means to achievement, the story of some forty years in the life of Irish Steel may be brought to a conclusion.

One need not strain prophetic confidence too far to suppose that before this century ends the steel company and its planners will be engaged in fresh projects for expansion, not, as in 1977, seeking the means to survive an economic storm, but instead providing essential additional capacity for a flourishing business. At that point in time, when a new generation observes as normality a well settled and prosperous industry, the first era of struggle and survival will doubtless have faded into a dim historical horizon, becoming almost a remote incident in Irish industrial evolution. Even so, the chronicle of these early years can still retain some value and entertainment for those who, long afterwards, find themselves curious to know how the steel industry began in Ireland; why it was started so late in the day; why on an island in Cork Harbour; and how the venture, attended by so many hazards and perplexities, still managed to lay solid foundations for eventual success.

Note on the Development Proposals, 1977 (see pp. 212–15)

Shortly after the Minister for Industry and Commerce, Justin Keating, had announced the government approval of the Irish Steel plans for modernisation and expansion of the industry, the chairman of the company held a press conference in Cork. The main features and objectives of the scheme are summarised in the following précis of Kevin McCourt's statement on 23 March 1977:

The Scheme aims at providing at Haulbowline a modern viable plant capable of manufacturing and marketing successfully a reasonable range of products in free competition within Western Europe and substantially export orientated.

Production capacity in terms of crude steel will be more than doubled, to a figure of 345,000 tonnes annually. The product range in finished bars and sections will be widened.

More than 50 per cent of the finished products will be sold in export markets. It is planned to establish sales agencies at strategic points within the EEC.

The basic raw material for steel making will continue to be cold-charged scrap. Annual consumption of scrap by ISH will increase to 393,000 tonnes of which about 274,000 tonnes will be imported. The new electric arc melting furnace will, however, be capable of utilising direct reduced iron when circumstances favour use of that raw material.

Plant To achieve the target outputs new and improved plant, equipment, buildings, transport and handling facilities are required. There will also be considerable changes in plant lay-out for more efficient operation.

The main new items (as distinct from extensions and improvements) are:

(1) An electric arc furnace of 90 tonne capacity. (The existing 30 tonne E.A. furnace will be kept in commission as reserve capacity.)
(2) A 3 strand continuous billet casting machine and ancillary equipment.
(3) A new overhead 150 t. crane to serve the melting shop.
(4) A new melting shop building to house (1) (2) and (3).
(5) A single modern rolling mill (to replace the two existing mills) with increased capacity and flexibility.
(6) A new billet reheating furnace of 77 t.p.h. capacity.
(7) Construction of a new despatch wharf.

(Note. The galvanising plant is not affected by the development plan.)

Cost The estimated cost of the project is £40 million, which includes £11.5 million of working capital.

Of the £28.5 million of fixed capital, not less than £8 million will be spent in Ireland.

The capital required to finance the project will be highly geared. The plan involves an equity subscription of £6 million, an IDA grant of £4 million, long-term credit of £24 million and short term borrowings of £6 million.

Timing It is anticipated that the project will take two years to complete from placing of contracts to commissioning of the new installations.

Employment A modern steel works is a capital intensive industry. The plan does not involve increased employment at Haulbowline. It aims to provide secure and permanent employment for the existing labour force, who with their families, comprise almost 3,000 people dependent on the success of the industry. At existing levels of remuneration, the Company pays some £3.5 million in wages and salaries each year.

National benefit Assuming export targets are reached, there will be a net gain to the balance of payments of not less than £10 million annually.

In addition to the national economic benefit of production and employment in a vital industry and the security of native steel supplies for industrial purposes, there will also be annual payments of £4.5 million to the ESB (Electricity Supply Board) and of £7 million for other Irish materials and services. The chairman ended his statement as follows:

> The industry will be profitable. It will service its capital and it will pay dividends on the state's investment. It will in addition be an international industry because of its exports. We must grow into a highly marketing-oriented company with a capacity to find outlets for our products and literally and metaphorically to deliver the goods to satisfied customers domestically as well as overseas.
>
> We take pride in our achievements in the past in the lower Cork Harbour area and we look forward with confidence and with a sense of responsibility to adding immeasurably to that record in the years ahead of us.

Index